COMPUTER SCIENCE, TECHNOLOGY AND APPLICATIONS

ADVANCES IN COMPUTATIONAL MODELING RESEARCH

THEORY, DEVELOPMENTS AND APPLICATIONS

COMPUTER SCIENCE, TECHNOLOGY AND APPLICATIONS

Additional books in this series can be found on Nova's website under the Series tab.

Additional e-books in this series can be found on Nova's website under the e-book tab.

COMPUTATIONAL MATHEMATICS AND ANALYSIS SERIES

Additional books in this series can be found on Nova's website under the Series tab.

Additional e-books in this series can be found on Nova's website under the e-book tab.

COMPUTER SCIENCE, TECHNOLOGY AND APPLICATIONS

ADVANCES IN COMPUTATIONAL MODELING RESEARCH

THEORY, DEVELOPMENTS AND APPLICATIONS

ANNA BELYA KORA
EDITOR

nova publishers
New York

For permission to use material from this book please contact us:
Telephone 631-231-7269; Fax 631-231-8175
Web Site: http://www.novapublishers.com

LIBRARY OF CONGRESS CATALOGING-IN-PUBLICATION DATA

Advances in computational modeling research : theory, developments and applications / [edited by] Anna Belya Kora.
pages cm
Includes bibliographical references and index.
ISBN 978-1-62618-065-9 (hardcover)
1. Mathematical models--Data processing. I. Kora, Anna Belya, 1971-
QA401.A276 2013
006.3'1--dc23

2013002539

Published by Nova Science Publishers, Inc. † New York

CONTENTS

PREFACE

Agent-based models (ABM) and computational models simulate the actions and interactions of autonomous agents (both individual or collective entities such as organizations or groups) with a view to assessing their effects on the system as a whole. They combine elements of game theory, complex systems, emergence, computational sociology, multi-agent systems, and evolutionary programming. In this book, the authors discuss the modeling, simulation and optimization of resources management in hospital emergency departments using the agent-based approach; computational modeling in male reproduction; consumer-centric agent-based marketing models; Lattice-Boltzmann simulation of transport phenomena in agroindustrial biosystems; agent-based models in cancer prevention research; model-based approaches to diagnosis of multi-agent plans; and modeling behavior of web users as agents with reason and sentiment.

Chapter 1 - Hospital emergency departments (EDs) are a primary healthcare unit, usually at the main entrance to the hospital and a key component of the whole healthcare system. The increasing demand of urgent care and the overcrowding of EDs are phenomena shared by health systems around the world. A naive solution could be to increase the size of EDs, but this is not a viable option due to the worldwide limitation of operational budgets.

Resource planning in EDs is complex because its activity varies depending on time, day of week and season. So the ability to manage both regular and special situations, such as seasonal or unexpected increases in ED demand, is the key for the efficient use of resources.

Simulation enables us to answer questions like “what happens if?” (e.g., in case of a specific pandemic, to explore how the composition of ED staff members influences the number of patients attended in a period of time) and to find the answer to questions such as “which is the best for ...?” (e.g., the number of healthcare staff that leads to minimizing the “Length of Stay” of patients, constrained by the availability of budget and number of healthcare staff), the optimization is needed.

Discrete Event Simulation (DES), System Dynamics (SD) and Agent-Based Modeling and Simulation (ABMS) are the main approaches used in the modeling and simulation of healthcare systems. In this chapter, a basic survey of these approaches, describing the characteristics of each one of them, including their pros and cons, allows us to show the advantages of the ABMS strategy.

The defined ED model is a pure Agent-Based Model, formed entirely by the rules governing the behavior of the individual agents which populate the system. Two distinct types of agents have been identified, active and passive. Active agents represent the persons involved in the ED (patients and ED staff), whereas passive agents represent services and other reactive systems, such as the information technology infrastructure or services used for performing tests.

Each agent (individual) is represented through a state machine. This includes all the required variables to represent the many different "states" in which an agent may be during the time it is in the ED. The changes of these variables, fired by inputs received from other agents, generate the states' transition. For modeling interaction between agents, the model includes communication between individuals and the physical environment in which agents interact (admissions, triage booths, waiting rooms, or consultation suites).

Using the simulation environment Netlogo, the simulation and optimization tools have been developed to be integrated in a Decision Support System (DDS), in order to help the managers to enhance the operation of an ED, providing additional knowledge for deciding on strategies such as patient admission scheduling, healthcare staff composition or clinical resources available. The DSS can be applied in different EDs, after a previous adjustment of the configuration parameters achieved through a tuning process.

This research has been developed in collaboration with staff members of the Parc Tauli Hospital in Sabadell.

Chapter 2 - In recent years we are seeing an increase in male infertility, with important negative consequences on our present and future life, such as population ageing, increase in expenses for National Health Services, worsening of sexual, personal, and social strains. Unfortunately, in about 50% of cases, it is impossible to perform an adequate diagnosis. This is likely due to the specific biology of male gametes. In fact spermatozoa gain their fertilizing ability only after they reside within the female genital tract, where important physical-chemical modifications, the "capacitation", occur. At present many aspects of this process are still unsatisfactory known. In the authors' opinion this could be the reductionist approach adopted until now in the study of this topic. Then, in consideration that spermatozoa are constituted by a network of heterogeneous components that interact collectively and nonlinearly, characterized by a complex behaviour in which the whole system is more than the sum of its single components, for the first time the authors adopted a computational modeling-based approach to study sperm capacitation. The authors realized the biological network of molecular event occurring during this process in Human sperm. As a result, in Chapter 2, the authors have found that spermatozoa are an ideal candidate to this approach because their molecular composition is stable, it is possible to manage them outside the organism, and it is possible to in vitro test the achievement of capacitation. Topological analyses of network demonstrated that capacitation is characterized by a high signaling efficiency and robustness against random failure. In a further experiment the authors modelized boar sperm capacitation, separating the molecules depending on their subcellular compartment. The analysis of model suggested that actin cytoskeleton is not only a mechanical support but it could exert a key role in coordinating the capacitation-related biochemical events. This hypothesis was, then, successfully validated by a traditional in vitro approach. More recently, they compared the models representing sperm activation in sea urchin, Caenorhabditis elegans and Human. It was found that these organisms share a similar signalling and metabolism architecture, reasonably because of the maintenance of ancestral mechanisms, characterized by important biological features such as robustness against random failure and signalling fastness, rapidity and efficiency.

In conclusion, from computational models it is possible to infer some important information not otherwise obtainable, and leading to achieve important goals, such as the understanding of complex interactions that characterize spermatozoa biology, potentially opening new perspective in drug discovery and diagnosis and therapy of male infertility.

Chapter 3 - The use of micro-simulations, and in particular the subset known as agent-based marketing models (ABMMs), are beginning to find a following in the marketing sciences. They have been used to model, for example, diffusion of innovations, marketing mix models, supply chain management issues and geospatial retail store plans. Consumer packaged goods companies, most notably Procter & Gamble, have also used ABMs to improve managerial decision making. Rand and Rust provide a summary, as of date, of other uses for ABMs in marketing. Despite the growing list of companies using agent-based models (ABMs) and marketing academics focusing on modeling consumer markets, ABMMs have still not become a common methodology used by marketing scientists.

In the first part of this chapter, the authors draw on the research of Leeflang and Wittink who laid out a foundation for building marketing models. They use their foundation to outline an approach for developing ABMs. In the second part of this chapter, they discuss details of the issues that arise in verification, validation and calibration for consumer-centric agent-based models, which is called agent-based marketing models (ABMMs). Although the focus of this chapter is on consumer-centric agent-based models, it is still relevant for any programmer looking to develop a validated model that is meant to be used to guide managerial decision making.

Chapter 4- Numerical methods are helpful either when mathematical difficulties prevent analytical solutions or when obtaining experimental data becomes inhibited due to safety, technological or economic issues. Computational modeling of food and bioprocesses has continuously increased as its importance has been recognized while suitable numerical techniques have been developed and applied. Chapter 4 is particularly devoted to lattice-Boltzmann method (LBM) simulation of transport phenomena involving agroindustrial biosystems.

One may distinguish LBM from both microscale simulation (e.g., molecular dynamics) and macroscale methods (e.g., finite-differences, finite-elements or finite-volumes) inasmuch as the bulk medium is treated like a collection of fictitious particles occupying and traveling within a discrete space, namely, a lattice. From the computational viewpoint, those particles basically follow two procedures, accomplished in discrete time steps. During "collision", particles bump into each other as they arrive at lattice sites and their velocities are rearranged accordingly. During "streaming", collision results are transported to adjacent sites as particles resume their motion along lattice links. By imposing conservation principles to such particle dynamics, macroscopic medium behavior can be numerically simulated.

This chapter particularly focuses on LBM fundamentals as well as on its application to heat and mass (species) transfer involving agroindustrial biosystems. LBM simulation is addressed for biospecific affinity chromatography, supercritical fluid extraction and blast-cooling of catering meals, which are among authors' current research work on computational modeling.

Chapter 5 - Agent-based models have proven very useful for studying complex dynamic processes. As biomedical researchers have turned their attention to the most complex diseases, they have made increasing use of agent-based models in various areas, including cancer biology. Cancer is arguably the ultimate complex biological system. It has recently become clear that much of its complexity arises through Darwinian evolution, driven by uncontrolled natural selection among rapidly mutating somatic cells.

In this chapter, the authors review the application of agent-based models and simulations to five key open questions in cancer prevention research. The five research questions they review as case studies include: 1) What normally prevents somatic evolution from generating cancer? 2) Why do benign growths often progress to malignant cancers? 3) What is the chronological sequence of molecular events in cancer progression? 4) Are there reliable molecular biomarkers for cancer?; and 5) Will evolved drug resistance stymie efforts at long-term cancer chemoprevention?

They conclude that molecular analysis can be usefully augmented by using agent-based models to generate plausible and productive hypotheses for empirical testing.

Chapter 6 comprises a comprehensive overview of major theoretical advances in agent-based modeling as well as a computer simulation of the Web 1.0 and Web 2.0 networks. Both of these networks are represented on the basis of the graph theory and are simulated using the Holistic-Emergent Social Interaction-Oriented Dynamics (HESIOD) model. This model enables a deeper understanding and interpretation of the dynamical features, stability and self-organization properties of the networks. Several suitable measures are employed to describe the state of the systems, including the Lyapunov exponent and information entropy. The results vividly indicate that what mainly characterizes Web 2.0, and radically differentiates it from Web 1.0, is a far-from-equilibrium dynamics with lower stability and self-organization, and higher diversity. As it is demonstrated with a real-world example, these characteristics can be attributed to the explosion of the user-generated online content, that is, a fundamental bottom-up process. This process surprisingly reinforces social dynamics and substantially offers the necessary conditions for the unpredictable creativity, innovation and adaptation.

Diagnosis is an essential step in the process of making robust complex systems and in particular Multi-Agent Systems (MASs). Chapter 7 reviews the most important model-based techniques to diagnosis of MASs that have been proposed over the last decade.

Since the term MAS refers to a wide range of systems, the Chapter focuses on MASs that can be represented as Multi-Agent Plans (MAPs), and categorizes the approaches to diagnosis of MAPs along different dimensions. Two features are particularly important to characterize the examined methodologies: the MAP model, which can either be implicit or explicit, and the notions of diagnosis (i.e., social, plan, or agent diagnosis). This categorization enables a direct comparison between the different solutions highlighting their advantages and disadvantages, as well as their similarities.

The chapter ends with some remarks about future developments on Model-Based Diagnosis (MBD) applied to MAPs.

Chapter 8 - A brief recapitulation of recent progress in modeling the collective emotional behavior of Web users is given based on the author's work. A comparative relation of the models devised for the online social networks, chats and blogs, reveals suitable structure of the model and minimal requirements for the agent's attributes. Furthermore, the character of the control parameters which can be deduced from high-resolution empirical data is discussed in Chapter 8. The potentials of the agent-based modeling of human dynamics are highlighted by few results. Some limitations of this modeling approach are also pointed out.

In: Advances in Computational Modeling Research
Editor: Anna Belya Kora
ISBN: 978-1-62618-065-9

Chapter 1

MODELING, SIMULATION AND OPTIMIZATION OF RESOURCES MANAGEMENT IN HOSPITAL EMERGENCY DEPARTMENTS USING THE AGENT-BASED APPROACH

***Manel Taboada*[1,*], *Eduardo Cabrera*[2], *Emilio Luque*[2], *Francisco Epelde*[3] *and M. Luisa Iglesias*[3]**

[1]Tomas Cerda Computer Science School, University Autonoma of Barcelona (UAB), Avenida de la Generalitat, Sant Cugat, Spain
[2]Computer Architecture and Operating Systems Dp., University Autonoma of Barcelona (UAB), Bellaterra, Spain
[3]Hospital of Sabadell, Corporació Sanitaria Parc Taulí, Sabadell, Spain

ABSTRACT

Hospital emergency departments (EDs) are a primary healthcare unit, usually at the main entrance to the hospital and a key component of the whole healthcare system. The increasing demand of urgent care and the overcrowding of EDs are phenomena shared by health systems around the world. A naive solution could be to increase the size of EDs, but this is not a viable option due to the worldwide limitation of operational budgets.

Resource planning in EDs is complex because its activity varies depending on time, day of week and season. So the ability to manage both regular and special situations, such as seasonal or unexpected increases in ED demand, is the key for the efficient use of resources.

Simulation enables us to answer questions like "what happens if?" (e.g., in case of a specific pandemic, to explore how the composition of ED staff members influences the number of patients attended in a period of time) and to find the answer to questions such as "which is the best for ...?" (e.g., the number of healthcare staff that leads to minimizing the "Length of Stay" of patients, constrained by the availability of budget and number of healthcare staff), the optimization is needed.

* E-mail address: manel.taboada@eug.es; manel.taboada@gmail.com. (Corresponding author)

Discrete Event Simulation (DES), System Dynamics (SD) and Agent-Based Modeling and Simulation (ABMS) are the main approaches used in the modeling and simulation of healthcare systems. A basic survey of these approaches, describing the characteristics of each one of them, including their pros and cons, allows us to show the advantages of the ABMS strategy.

The defined ED model is a pure Agent-Based Model, formed entirely by the rules governing the behavior of the individual agents which populate the system. Two distinct types of agents have been identified, active and passive. Active agents represent the persons involved in the ED (patients and ED staff), whereas passive agents represent services and other reactive systems, such as the information technology infrastructure or services used for performing tests.

Each agent (individual) is represented through a state machine. This includes all the required variables to represent the many different "states" in which an agent may be during the time it is in the ED. The changes of these variables, fired by inputs received from other agents, generate the states' transition. For modeling interaction between agents, the model includes communication between individuals and the physical environment in which agents interact (admissions, triage booths, waiting rooms, or consultation suites).

Using the simulation environment Netlogo, the simulation and optimization tools have been developed to be integrated in a Decision Support System (DDS), in order to help the managers to enhance the operation of an ED, providing additional knowledge for deciding on strategies such as patient admission scheduling, healthcare staff composition or clinical resources available. The DSS can be applied in different EDs, after a previous adjustment of the configuration parameters achieved through a tuning process.

This research has been developed in collaboration with staff members of the Parc Tauli Hospital in Sabadell.

1. Introduction

Emergency departments (ED) may well be one of the most complex and fluid healthcare systems that exist, receiving an increasing demand and usually being overcrowded. The resource planning in the ED is complex because its activity is not linear, and it varies depending on time, day of week and season. For this reason the ability to simulate special situations such as seasonal increasing of the ED demand can be useful for the efficient use of resources.

There are no standard models to describe complex systems, but simulation becomes an important tool for their modeling. Discrete event simulation (DES), system dynamics (SD) and agent-based modeling and simulation (ABMS) are the three main approaches used when simulating healthcare systems. There is a large body of literature describing the use of DES models in ED studies, whilst there is considerably less literature on the use of ABMS for this purpose, although healthcare systems are based on human actions and interactions that are very difficult to model with DES, and can be more properly modeled with ABMS.

This chapter presents the results of an ongoing project that is being carried out by researchers of the group High Performance Computing for Efficient Applications and Simulation (HPC4EAS) of the Department of Computer Architecture and Operating Systems of the Universitat Autonoma de Barcelona (UAB), in collaboration with the team at the Emergency Services Unit at Hospital de Sabadell (Parc Tauli Healthcare Corporation), a hospital that provides healthcare to a catchment area of 500,000 people attending 160,000 patients/year in the ED. The objective of such project is to develop a model and a simulation that, used as decision support system (DSS), provides the ED Directors with additional knowledge of patient admission scheduling, physician staff, and alternatives ways for decreasing patients' length of stay (LoS), amongst other situations, but also aids them in setting up strategies and management guidelines to achieve resource optimization and to enhance the efficiency of the ED.

The optimization of the number of staff members of EDs (including admission staff, triage nurses, and doctors), is a multidimensional problem. Such problem has the following equation (1):

$$\begin{aligned} &\max / \min f(x) \\ &\text{subject to } x \in C \end{aligned} \quad (1)$$

where x is the variable, f is a function ($f : C \rightarrow R$), C is the constraint set, and $\exists X_o \in C$ such that $f(x_o) \leq f(x)\ \forall x \in C$ for minimization, and $f(x_o) \geq f(x)\ \forall x \in C$ for its counterpart, maximization. x is discrete, global sampling from x is possible, and f (x) at any $x \in C$ cannot be evaluated exactly, and for this reason it must be estimated via a simulation procedure. In the specific case of the ED the goal is to identify the mix of staff members (number and level of experience) that let to optimize the performance of the ED, which due to the complexity of the system, needs to be estimated via a simulation. Optimization via simulation is a difficult problem [7]. Only few alternatives might be explored because estimating f (x) at a single point $x \in C$ in equation (1) may require substantial effort, and in fact, simulations are usually computationally expensive.

The work has been developed following an iterative and spiral methodology, divided in cycles of five phases:

1. System analysis: the information was obtained from two different hospitals. The first one was the Hospital of Sabadell, a large center, and the second a medium size hospital, the Hospital of Mataró, that provides care service to a geographical area with a population of 250,000 people, attending an average of 110,000 patients per year in its ED. Such information was collected such the following tools:
   - Focus groups with the directors of the ED, the coordinators of the main areas of the ED, and different kinds of the ED staff (admission personal, triage nurses, sanitarian nurses and doctors).
   - Observation of the different ED zones of both Hospitals, with the purpose of analyzing and taking notes about how the different processes take place.
   - Data given by the Information System Department of the Hospitals (ISDH), concerning to the activity carried out by the ED.

   As result of this phase the key information of the system could be identified (different kinds of agents, the interactions that take place, the different areas, etc).
2. Model design: This consists of the formal definition of the model, including the environment, the different kinds of agents, and their behavior.
3. Simulator implementation: the simulation of the model has been implemented using the agent-based simulation environment NetLogo [27], a high level platform that lets modelers give instructions to hundreds or thousands of independent agents, all

operating concurrently, making possible to explore connection between the micro-level behavior of individuals and the macro-level patterns that emerge from the interaction of many individuals.

4. Simulator execution and results analysis: introduction in the simulator the configuration and input data given by the ISDH, then the execution of the simulation to obtain the output data that enables us to carry out the next phase.
5. Simulator validation: this phase is exhaustively explained in section 3.7.

Once the cycle has been completed, and based on the conclusions obtained during the analysis and validation phase, the model is updated and a next cycle is started. The process is repeated until the objectives are achieved.

After completing the second cycle, an Agent-Based Model for Hospital Emergency Departments has been obtained, in which the system behavior emerges as a result of agent's actions and interactions. This model describes the complex dynamics found in an ED, representing each individual and system as an individual agent. Two distinct kinds of agents have been identified, active (the persons involved in the ED such as patients and ED Staff) and passive (reactive systems, such as the information technology (IT) infrastructure or labs that perform tests).

State machines are used to represent the actions of each kind of agents. This takes into account all the variables that are required to represent the different states in which such agent may be throughout the course of time in ED. The change in these variables, invoked by an input received from an external source, is modeled as a transition between states. In order to control the interactions that take place between agents, both the communication between individuals as well as the physical environment in which these individuals interact is modeled.

The remainder of the chapter is organized as follows; section 2 describes the previous work. The proposed emergency department model, the simulator features and the details of the verification and validation methodology are detailed in section 3. The specific issue of optimization is explained in section 4, including the results of the simulation optimization experiment and the proposed reduced scheme. Finally, section 5 closes the chapter with conclusions and future work.

2. Literature Review

Simulation modeling approach started to be used to solve healthcare problems in the US and UK more than three decades ago. In 1979 computer simulation was applied to hospital systems to improve the scheduling of staff members [9], and in Saunders et al. [23] the aim was to quantify the impact that the amount of staff members and beds had on patient throughput time. Moreover, a survey of discrete-event simulation in healthcare clinics was presented in Jun et al. [13]. Especially over the last decade there have been fruitful efforts in developing simulation-optimization models for solving healthcare management problems [1, 26].

Following Brailsford [4], and taking into account the role of different system modeling tools in study of healthcare systems, the models may be classified into three groups:

1. Disease models: often used for studying the clinical effectiveness or cost-effectiveness of some intervention;
2. Operational or tactical models at the healthcare unit level: used for capacity planning, resource allocation and process redesign;
3. And finally strategic or system-wide models, which often do not model individual patients at all.

Discrete event simulation (DES), system dynamics (SD) and agent-based modeling and simulation (ABMS) are the main three approaches used in simulation of healthcare systems. There is a large and growing body of literature describing the use of DES and SD models in Emergency Department studies. However there are few reports of the use of ABMS for this purpose, although healthcare systems are based on human actions and interactions, and the ability of ABMS to represent this makes it an appealing approach, while DES is not well suited to model them [5]. The potential of the use of agent-based modeling and simulations techniques in emergency departments modeling is clearly illustrated and shown in Kanagarajah et al. [14], Günal [8] and Staindsby et al. [24].

Modeling techniques using agents can bring the most benefit when they are applied in human systems in which agents exhibit complex and stochastic behavior and the interaction between agents are heterogeneous and complex [3]. In the particular case of social sciences ABMS is used in situations in which human behavior cannot be predicted using classical methods such as qualitative or statistical analysis [17].

Paulussen et al. [18] describe a multi-agent based approach to patient scheduling in Hospitals. In such a system patients and hospital resources are implemented as autonomous agents in which the resource agents only see patients as entities to be treated, and the patient agents only see the medical actions as tasks that need to be performed. The coordination of patients is achieved through a market mechanism. Patient agents negotiate with each other over scarce hospital resources, using state health dependent cost functions to compute bid and ask prices for time slots. Within this concept, stochastic processing times and variable pathways are considered. Unfortunately the system doesn't take into consideration patient variety or the different kinds of healthcare staff.

In 2008 Hutzschenreuter et al. [10] present an agent-based simulation and evaluation tool for patient admission scheduling, with the aim of achieving an efficient use of the hospital resources through the combination of different profiles of resource use. The proposed system includes multiple patient groups with stochastic arrival and treatment pathways. Authors also present an approach to optimize resource management using the model with brute-force. Again the model focuses on hospital domain, specifically in cardiothoracic surgery, and considers 4 different profiles of patients. Such patients are only differenced by their Length of Stay type and their distribution of arrival, which means that the patient's behavior is not actually modeled. The different hospital units are considered as agents. For this kind of agent the model considers the amount of resources, the cost, the resources availability and the scheduling policy.

The work carried out by Jones et al. [12] is a specific example of simulation applied to Hospital Emergency Departments. Authors present an agent based simulation tool, developed with NetLogo, and designed to evaluate the impact of various physician staffing configurations on patient waiting times. One of the limitations of the model is that it only considers physicians and patients, missing the other ED staff. In addition, both the programmed rules for the physicians and the patients' model are too simple, and it only takes into account 3 characteristics: acuity, discharge position, and length of service time. These simplifications become the non-realistic model. Moreover, the work only focus on door-to-doc time (the time between the patients' arrival to the ED, and when they are called by a physician), but it has not taken into account what happens during the "diagnostic and treatment process".

One of the few works that has been carried out with real data is Laskowski et al. [16]. The authors describe an agent based model of an emergency department and its utility for evaluating workflow and assessing patient diversion policies. The simulation includes multiple emergency departments in order to evaluate redirection policies of ambulances or patients. The proposed agent-based model allows simulating a number of emergency departments and introduces a method of extracting real time patient data from emergency departments throughout a city, with the overall objective of evaluating alternative patient-diversion policies.

There is another work in which Netlogo is used to simulate the workflow of an emergency department [25]. This work introduces an agent-based simulation which models various phases in ED workflow, such as triage, nurse screen, resident exam, attending exam, lab/radiology, and disposition. Unfortunately, real data is not used, and the acuity of patients is not considered. In the words of the authors, the workflow and rules that control the agents' behavior are very simple and not realistic, and it is not applicable to different ED.

In some works simulation optimization is used to improve the operation of ED. Ruohonen et al. [20] use a commercial simulation package for developing a simulation model which describes the operations in the Emergency Department of Special Health Care at the Central Hospital of Jyväskylä, Finland. The tool was used to test different process scenarios, allocation of resources and perform activity based cost analysis. By using the simulation model authors identify a new operational method (Triage-team method) that permits improving the effectiveness of the Emergency Department of Special Health Care. Although it is not an Agent-Based Model, it includes the different units of the ED (trauma, surgery, neurology, internal medicine and Children's medicine), 3 different entities (patients, blood tests samples and laboratories test results), and 5 different resources (Secretary, Nurse, Doctor, Lab Staff and X-ray Staff). One of the main limitations is that the model doesn't consider different kinds of patients or ED staff (in terms of level of experience).

Ahmed et al. [1] design a decision support system, based on the Discrete Event Simulation approach, for the operation of an emergency department unit in a governmental hospital in Kuwait. They present a methodology that uses system simulation combined with optimization to determine the optimal number of staff members required to maximize patient throughput and to reduce patient time in the system subject to budget restrictions. Such methodology involves a complex stochastic objective function under to deterministic and stochastic set of restrictions.

An evolutionary multi-objective optimization approach is used for dynamic allocation of resources in hospital practice in the work carried out by Hutzschenreuter et al. [11] referenced above, while Persson et al. [19] found that combining agent-based approaches and classical optimization techniques complement each other. Specifically these authors compare the strengths and weaknesses of *agent-based and classical optimization approaches* and evaluate their appropriateness for the specific case of the dynamic distributed resource allocation. With the purpose of finding approaches capitalizing on the strengths of both of them, they carried out some experiments that permit identifying two promising hybrid approaches: using coarse planning and embedded optimization techniques

The model presented in this chapter is a specific case of *operational model.* The proposal addresses many of the issues surrounding the modeling and simulation of a hospital emergency department using agent-based paradigm. In an attempt to understand micro-level behavior the basic rules governing the actions of the individual agents are defined. Macro-level behavior, which means the system as a whole, emerges as a result of the actions of these basic building blocks, from which an understanding of the reasons of system-level behavior can be derived.

Compared with previous work, the model presented in this chapter offers two main remarkable improvements. On one hand, the generality of the model, which means the possibility of applying model and simulation in different hospital ED, after a previous adjustment of the configuration parameters achieved through the tuning process that is described in section 3.7. On the other hand, the variety and typology of agents are richer. The variety of patients in this model is higher and more realistic that the one found in the models of previous works. In addition, the model includes ED Staff (admission staff, sanitarian technicians, nurses, doctors, etc) and devices (Information System Infrastructure, radiology services, laboratory tests, etc) that aren't considered in other consulted works. Moreover, in the case of the ED staff, it takes into account different levels of expertise. Undoubtedly, all these elements may affect the patients' LoS, thus in the performance of the emergency department.

3. Agent Based Model of the Emergency Department

As explained beforehand, the emergency department model has been designed through the information obtained during the round of interviews carried out with ED staff at the Hospital of Mataro and the Hospital of Sabadell. It is a pure Agent-Based Model, formed entirely of the rules governing the behavior of the individual agents which populate an ED. Two kinds of agents have been identified, active and passive. *Active agents* represent people, who act upon their own initiative:

1. *Patients.*
2. *Admission staff*: who receive patients just when they arrive at the ED.
3. *Healthcare technicians*: who help certain patients to move from one place to another of the ED.
4. *Triage and emergency nurses*: the former receive patients after their admission for establishing their priority level. The second involved in the diagnosis and treatment phase.
5. And *doctors*.

Passive agents represent systems that are solely reactive, such as the loudspeaker system (that is used by ED Staff to communicate with patients that are in the waiting rooms), patient information system, pneumatic pipes (that are used to send the trials from the diagnostic zone to labs), and central diagnostic services (radiology service and laboratories).

The ED is divided into different zones in which different types of agents may act, maintaining interactions that also may be different. Such interactions are carried out through communication. For this reason the model includes both the environment and a communication model.

The various components of the general model are described in detail in this section. Sections 3.1 and 3.2 explain the manner in which active agents and passive agents, respectively, have been modeled. State variables are discussed in section 3.3. The communication model is defined in section 3.4 and the details of the environment are outlined in section 3.5. Section 3.6 presents information about the simulation implementation, and finally the model verification and validation is explained in section 3.7.

3.1. Active Agents

With the purpose of considering the most significant features of agents (interaction with the environment and other agents, their internal situation, their evolution over time, etc), the behavior of active agents has been modeled using *Moore State Machines* [15]. The Agent will remain in a specific state until, through interaction with other agents, they receive an Input (an Output generated by other agents) which causes a change in the state of such agent, generating an Output sent to the agent with whom they are interacting. As example (Figure 1), when the patient is waiting for being examined by the physician (initial state of the patient), upon receiving a question from the doctor (which is an Input received for patient, and an Output generated by the physician), he changes his state, and proceeds to answer the question (new state) providing the required information (Output generated by the patient, Input received by the physician).

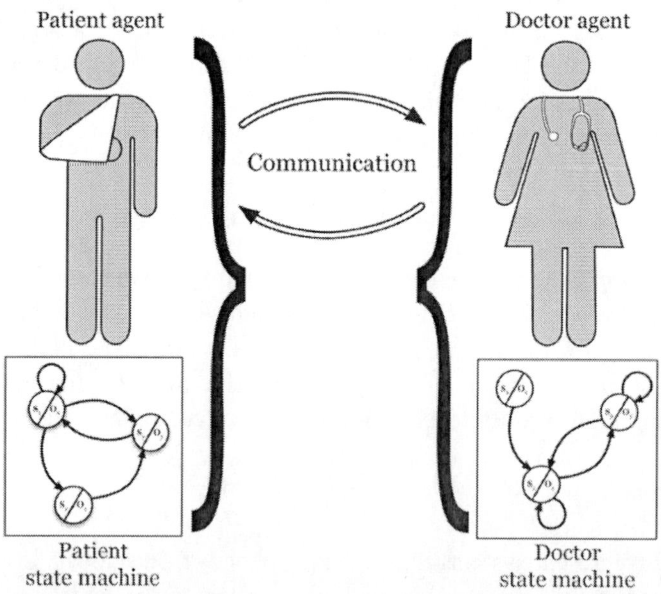

Figure 1. Formal representation, using state machines, of the interaction carried out between patient and physician during the exploration phase.

The agent's state machine will move to the next state (S_{t+1}) following the transition, which may be another state or the same one in which agent was before the transition. The input may be more accurately described as an input vector (I) that contains a number of input variables, each one of which may take a number of different values. As this is a Moore machine, the output only depends on the state, so each state has its own output, although various states may have outputs that are identical. Again, the output is more accurately described as an output vector (O), a collection of output variables, each with a number of defined possible values. Transitions between states depend on the current state at time t (St) and the input at time t (It).

In complex systems such EDs the model is not entirely deterministic. In this case, given a current state and an input combination, the state machine can be modeled with more than one possible next state. Each one of the input variables of the input vector may take a number of different values, and the state transition is defined with probabilities on the "effect" of the input. The question of which transition is made is chosen randomly at the time of the transition.

The state machine can be represented as a diagram (figure 2) but also in form of a state transition table (figure 3). The diagram shows the current state, S_X, that may receive the input Ia and move to either state Sy, state Sz, or remain in the same state, with a probability of p_1, p_2, and p_3 respectively. One of these transitions will always occur, which is to say $p_1 + p_2 + p_3 = 1$.

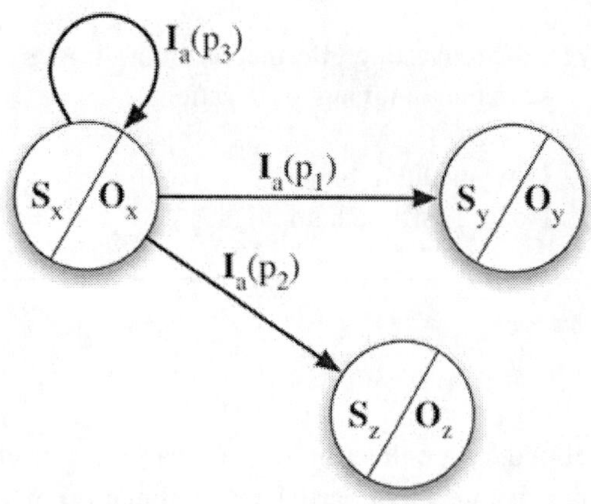

Figure 2. Probabilistic state transition diagram.

In the case of the table each row represents a unique state-input combination, showing the output and the state in the next time step (defined by the current state and the input).

3.2. Passive Agents

Passive agents represent services within the hospital system such as the information technology infrastructure (which allows storing patient details), radiology services and other laboratory tests, as well as specialist systems such as the pneumatic tube networks that some larger hospitals use to transfer samples from one part of a building to another quickly. In some of these kinds of agents the state machine will be a simple system for interacting with active agents. In order to represent data storage or other systems that may have a very large number of combinational states, a simple memory model will be used. A passive agent may (although it is not necessary) have a simple record based memory system, allowing it to store and repeat information provided by active agents.

Current state / Output	Input	Next state / Output
--	--	--
S_x / O_x	$I_{a(p_1)}$	S_y / O_y
S_x / O_x	$I_{a(p_2)}$	S_z / O_z
S_x / O_x	$I_{a(p_3)}$	S_x / O_x
--	--	--

Figure 3. Probabilistic state transition table.

Table 1. Initial selection of state variables and their values

Variables	Values
Name/identifier <id>	Unique per agent
Location <location>	Entrance; admissions waiting zone, admissions desk "X" ; waiting room "X"; triage box "X"; Consultancy Room "X"; Treatment Booth "X"; ...
Action	Idle; requesting information from <id>; giving information to <id>; searching; moving to <location>
Acuity level	I; II; III; IV; V
Communication skills	Low; medium; high
Level of experience (ED Staff)	None; Low; medium; high.

3.3. State Variables

The agent's state is defined by a collection of state variables, each with different possible values, so that each state corresponds to a particular combination of values of these variables. The transition will occur after receiving an input that causes a change in the value of one or more of these variables. Table 1 shows the set of state variables (and some of their possible values) that have been defined through a round of interviews, based on the minimum amount of information required to model each patient and member of staff.

Thus the initial set of state variables is:

- Name/identifier: for identifying each individual
- Location: for identifying the area of the ED where the individual is at each moment. The kind of active agents (specifically the ED Staff) who may be present and the kind of interactions that may occur are different in each ED area.
- Action: what the individual is doing in a particular state. The actions will be different depending on the agent and the place where such an agent is. Actions normally consume a specific amount of time, which will influence the patient's total Length of Stay in the emergency department.

- Acuity level: depending on the physical condition and the symptoms of patients, they are classified in 5 different levels of acuity and priority, from I (the maximum level) to V (the minimum), following the acuity scale used in Spanish Emergency Services [6].
- Communication skills: the time spent during a specific process will depend on the agent's ability to communicate. This variable has sense for both patients and ED staff.
- Level of experience: this only has sense for ED Staff. The time required by the Staff to complete a specific process will depend on their experience. The higher the experience is, the lower the time required will be.

3.4. Communication Model

The interaction between agents is carried out through communication, which is modeled as the input that agents receive and the outputs that they produce. The communication model represents three basic types of communication:

1. *1-to-1:* between two individuals. In this type the message has a single source and a single destination, as happens between admission staff and patient, during the admission process.
2. *1-to-n:* this represents an individual addressing a message to a group, like a doctor giving information to the patient and a nurse during the diagnostic process.
3. and *1-to-location:* when an individual speaks to all occupants of a specific area, for instance when any staff member uses the speaker system to address a message to all the people who are in a specific waiting room.

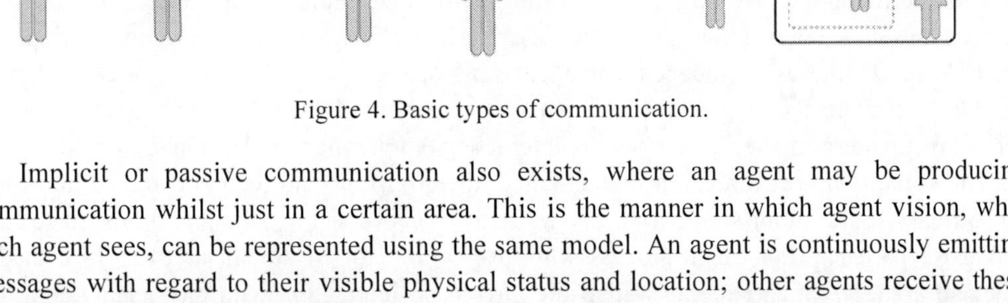

Figure 4. Basic types of communication.

Implicit or passive communication also exists, where an agent may be producing communication whilst just in a certain area. This is the manner in which agent vision, what each agent sees, can be represented using the same model. An agent is continuously emitting messages with regard to their visible physical status and location; other agents receive these *1-to-location* messages and may act upon them in certain circumstances. For instance an agent waiting for another agent in a certain area will receive communication that the agent has entered and act upon it, representing, for example, a nurse seeing a patient enter a triage room and attending them.

3.5. Environment

The ED is divided into different zones in which different types of agents may act, maintaining interactions that also may be different. For this reason the model includes the environment, but using a low level of positional precision, because it is enough to identify the areas and the distance between them, to take into account the time required to move from one to another.

The areas identified through the information obtained during the interviews carried out are:

- Admissions: where patients have to be addressed just when they arrive to the emergency department, and Admission Staff will register their arrival.
- Triage Booth: where a “triage nurse” will receive the patient, will take their vital signs and obtain some additional information in order to identify the priority level with which patients will be attended.
- Waiting Rooms (WR): there are two different waiting rooms. In one of them (WR 1) patients and their companions will be waiting for being triaged, after the admission process. In the second one (WR 2) patient will remain after the triage process, waiting to be called for being attended in the diagnostic and treatment zone.
- Diagnostic and treatment zone: where doctors, nurses and other ED Staff will carry out the diagnostic and treatment process with patients.

Figure 5 shows a representation of topographical distribution of the emergency department. Certain locations may be physically distinct but functionally identical, for instance the triage rooms or the treatment zone booths. The Simplified ED layout is distinct in order to represent that each available room may only be used by one ED Staff-patient group at a time. The environment also contains representations of the relative distances between different discrete locations.

3.6. Simulation Implementation

The simulation of the proposed model has been implemented using the agent-based simulation environment NetLogo, a high level platform particularly well-suited for modeling complex systems developing over time [2]. NetLogo lets modelers give instructions to hundreds or thousands of independent agents, all operating concurrently, which makes it possible to explore the connection between the micro-level behavior of individuals and the macro-level patterns that emerge from the interaction among many individuals.

The version of the simulation implemented after finishing the second cycle includes the four primary areas: admissions, triage, 3 waiting rooms (one for patients before triage, the second for patients after triage process who are waiting for treatment, and the third in the treatment area, in which patients wait while having no interaction with physicians), and the diagnosis and treatment area. The kind of active agents represented in this simulation are patients (P), admission staff (AS), triage nurses (TN), and doctors (D). In the specific case of the ED Staff, two distinct levels of experience have been considered (low, labeled as junior, and high, labeled as senior).

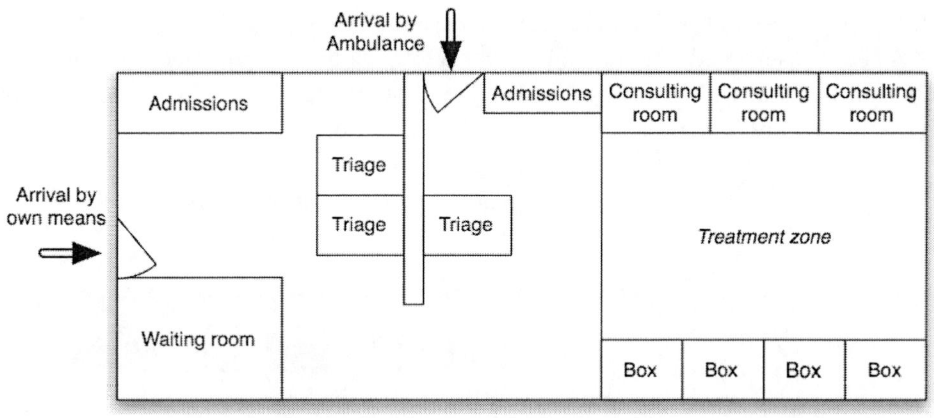

Figure 5. Simplified emergency department layout.

Depending on physical condition and symptoms of patients, through the triage process they are classified in 5 different levels of acuity and priority (from I, the maximum level, to V, the minimum) following the Spanish Triage Model, the acuity scale used in the Spanish Emergency Services. The diagnosis and treatment process will be different depending on the priority level. In addition, in accordance with the organization model used in large Spanish EDs, the treatment zone is divided into two different areas: one for treatment of patients with a level IV or V, and the other one for the attention of patients with a level I, II or III.

In the simulation implemented patients arrive at the ED by their own means, and wait to be attended in the admission zone. The simulation reads the input information from a text file within the data given by the Hospital within the number and typology of patients arriving at any time within a 24-hour cycle. Once the admission process has been carried out, the patient waits in a first waiting room (WR) until he/she is called by a triage nurse, who takes his/her vital signs and obtains some additional information in order to identify the priority level with which patient will be attended. After the triage process, the patient goes to a different WR and waits there until a doctor is free and calls him to start the process of diagnosis and treatment. Once this process has been completed, the patient leaves the ED. In the current version of the simulation patients of all the kinds (I to V) are considered until the triage process. But only the part of the diagnostic and treatment zone in which patients with a priority level IV or V are attended is considered.

The diagnostic and treatment process for such patients is divided into three phases:

1. Initial evaluation: a phase that takes 20 minutes on average, during which patient and physician remain in a treatment booth interacting;
2. Additional testing: during the first 10 minutes the patient and physician interact in the treatment booth, while the test is carried out. After that, the patient leaves the booth and waits in the WR until the results are sent. The total duration of this phase will depend on the Lab's workload;
3. Application of medication or treatment: patient is called to enter a treatment booth, and during the first five minutes the physician applies the treatment to the patient. Once the treatment has been applied, the patient leaves the booth and will be waiting in the WR until the treatment has taken its effects, which on average takes 45 minutes.

Considering the information given for the Hospital of Sabadell, 20% of patients attended in this zone are discharged after the initial evaluation, the 60% require additional testing, and only 20% of the total will remain for medication. The simulator takes into account all theses specifications. Figure 6 shows a screenshot of the simulation.

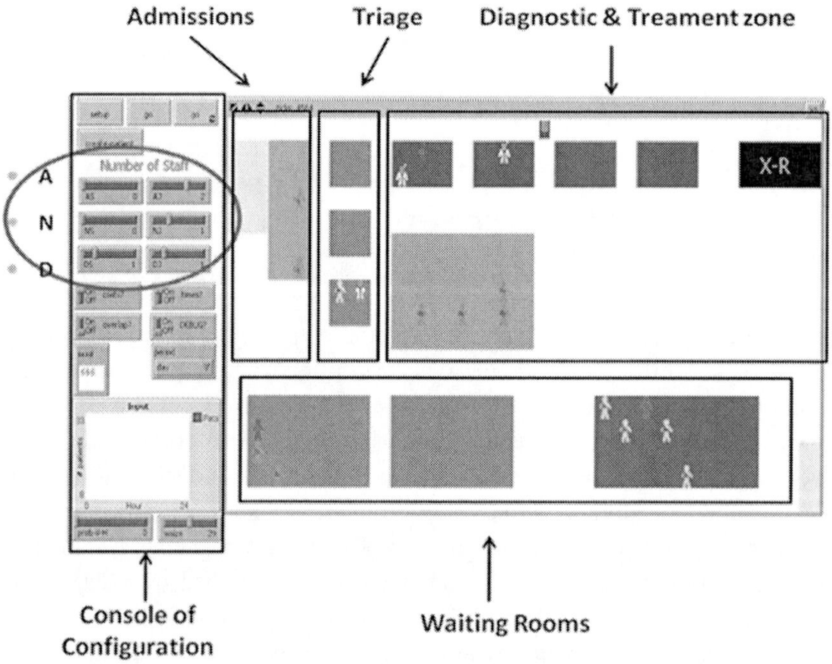

Figure 6. Screenshot of Netlogo with the features of the experiment.

The layout of the different parts of the ED implemented is shown in the central framework. The user can set both the number of each one of the different types of the ED staff (AS, TN and D), and their level of experience (junior or senior), through the configuration console which is on the left-hand side of the interface.

3.7. Model Verification and Validation

Taking the definitions of Sargent [22], model verification is the task of ensuring that the model behaves as intended, while model validation is the process of determining whether the simulation model is a useful or reasonable representation of the real system. With verification, the logic of the proposed model is examined, observing overall operations in the simulation ED and tracking different types of patients. In the case of validation, the most definitive method is to compare the output data from the simulation with the actual data from the existing system using formal statistical analyses such as confidence intervals.

Following the recommendations of Sargent, each one of the model iterations verification has been carried out performing the face validity on the conceptual model and exploring the simulation model's behavior. In the specific case of the validation, it only has been developed partially because the current version of the simulation is not the final one. In both cases the process has been carried out with the participation of the team from the Hospital of Sabadell

ED Staff, and it leads us to conclude that both model and simulation represent an ED correctly. The techniques applied are:

- Animation: the model's operational behavior has been displayed graphically. The movements of the different parts of the ED are shown graphically during the simulation run.
- Degenerate Tests: the logic of the proposed model has been examined through a varied set of experiments detailed below, observing overall operations in the simulation ED and tracking different types of patients.
- Face validity: the Hospital of Sabadell ED Staff has been heavily involved in all the process, verifying that the model behaves like the real system.

Set of Experiments Carried Out

In order to analyze the behavior of the simulator in front of the variables that can have an influence in the ED, the simulation has been executed several times for the equivalent to one day of activity. In a first exercise the amount of patients who arrive at the ED has been changed, considering four different scenarios in terms of arrival of patients per day: 93, 190, 284 and 397. The mix of ED Staff has been maintained invariable (1 AS, 2 TN and 4 D, all of them with a junior profile). Table 2 shows the results obtained. As expected, with the increase of the patients' arrival, if the available resources don't change, the system becomes saturated, which means a gradual increase in Patient's Length of Stay, in the waiting times at the WR1 and WR2, and in the number of patients waiting to be attended at the end of execution. In the first 2 scenarios, patients are attended without having to queue and there are idle resources, which means staff members who do not attend any patient during the execution. In the third scenario, some waiting time starts to emerge, although it is insignificant. By contrast, in the scenario of maximum patient arrival the average waiting time in WR1 is nearly an hour, and 2.44 in WR2, and there are a significant number of patients who remain in the waiting room without being treated at the end of the execution.

Table 2. Results obtained with an invariable mix of ED Staff, for an increasing arrival of patients

A - Total patients arrived to the ED	93	190	284	397
B - Total patients who completed the process during the execution	90	184	271	270
For patients that have completed the process:				
C - Average of total time of Stay in ED (mesured in hours)	0.54	0.54	1.08	3.95
D - Average of total time of stay in WR 1 (mesured in hours)	0	0	0.04	0.98
E - Average of total time of stay in WR 2 (mesured in hours)	0	0	0.52	2.44
F - Patients in WR 1 when execution finalizes	0	0	1	46
G - Patients in WR 2 when execution finalizes	0	0	8	74

A second exercise was carried out to analyze the effects caused by the change in the level of experience of ED staff over the throughput of the system. To do this both the patient's arrival and the number of ED staff have been kept invariable, considering 1 admission staff, 1 triage nurse an 2 doctors. A total of 12 executions were carried out, one for each of the possible combinations by changing the level of experience. The results are shown in table 3.

Table 3. Results obtained with a variable experience of ED Staff, for an invariable arrival of patients

	AS		TN		D								
	Jr	Sr	Jr	Sr	Jr	Sr	A	B	C	D	E	F	G
1	1		1		2		397	135	7.8	4.83	2.45	222	37
2	1		1		1	1	411	157	7.35	5.94	0.93	236	14
3	1		1			2	384	172	6.71	6.27	0	209	0
4	1			1	2		404	136	8.26	2.72	5.4	160	105
5	1			1	1	1	394	159	7.29	2.9	3.96	129	82
6	1			1		2	409	182	6.82	3.55	2.87	165	59
7		1	1		2		400	135	8.17	5.21	2.44	225	38
8		1	1		1	1	382	157	7.06	5.67	0.93	207	14
9		1	1			2	397	172	6.95	6.49	0	222	0
10		1		1	2		420	136	8.38	2.85	5.06	176	105
11		1		1	1	1	389	159	7.33	2.94	3.96	145	82
12		1		1		2	404	182	6.52	3.27	2.87	160	59

First column identifies the execution, the following 6 columns inform of the experience of the ED staff (the lowest level identified as Junior, and the highest one as Senior), and the last 7 columns (A to G) correspond to the results, using the same kind of information detailed in the first column of Table 2 (A-patients who arrived to the ED; B- patients who have completed the process during the execution; etc.). Considering the total number of patients and the time of stay in the ED, as it was expected the combination of worse outcomes is that in which all staff have a junior profile (execution 1, with a total of 135 patients attended, and an average time of stay in the service of 7.8 hours) and the best one is that in which all the staff have a senior profile (execution 12, with 182 patients and 6.52 hours), representing a 37% increase in treated patients and a 16.02% reduction in the attention time. Analysis of the data also leads us to conclude that the system is more sensitive to changes in the doctors' experience that in the rest of the ED staff. Finally, to reduce the average waiting time and the number of patients staying in the WR1, the experience of triage nurses should be improved, without changing the admissions staff. The same goes for the WR2, but in this case improving the experience of doctors without altering the experience of the triage nurses.

A third experiment was done in order to analyze the effects caused by the change in the number of ED staff over the throughput of the system. In this case the level of experience of ED staff stays invariable. A total of 11 executions were carruied out, one for each of the possible combinations by changing the number of staff members, taking into account that the staff number in one ED area should be equal or bigger than the number in the previous area. The results are shown in table 4, whose structure is similar to table 3. The worst combination is the execution with the lowest number of professionals (execution 1, with a total of 135 patients attended and an average time of 7.8 hours) and the best one is that with the maximum number of staff members (execution 11, with 271 patients and 3.8 hours). The increase of 5 professionals (1 AS, 2 TN and 2 D), generates a 100% increase in the number of treated patients, and a 51.28% reduction in the average time of stay. Due to the relative increase is higher in the staff of triage area, the results improve in WR1 (waiting time is zero, and there are no patient at the end of execution), but worse in the WR2 (number of patients waiting increase from 37 to 128, and the average waiting time from 2.45 hours to 3.27).

Executions 2 and 7 give similar results to those obtained in 1, but using a bigger number of professionals. In executions 4, 6 and 9 the number of treated patients is the same as in execution 11, but using fewer professionals. However, the average time of stay in the ED is higher (9 minutes in executions 4 and 9, and 32 minutes in the implementation 6). Considering these results, and the saving of resources, it can be concluded that execution 4 is the most optimal combination of the 11.

Considering the results obtained in these 3 experiments, it can be concluded that the gradual increase in the number of ED staff and the improvement of their experience bring about a gradual growth in the number of patients treated. In addition, to minimize waiting times in the different WR, a balanced distribution of both the number of different kinds of ED staff and their level of experience is important. Therefore, staff shortages can be partly solved by involving staff of higher experience.

Finally the model and simulation have been used for evaluating the effects over the ED performance (LoS and number of patients attended) of the derivation to ambulatory and primary care services of patients who go to emergency service without requiring urgent attention [21]. The derivation will be done after the triage process, the phase in which the priority level of patients is identified.

Table 4. Results obtained with a variable number of ED Staff, for an invariable arrival of patients

	AS	TN	D	A	B	C	D	E	F	G
1	1	1	2	397	135	7.8	4.8	2.5	222	37
2	1	2	2	399	135	8.2	0.7	7	48	212
3	1	2	3	398	203	5.8	0.6	4.7	48	143
4	1	2	4	397	271	4	1	2.4	46	75
5	1	3	3	397	203	6.1	0	5.6	0	188
6	1	3	4	406	271	4.3	0	3.8	0	128
7	2	2	2	382	135	7.6	0.3	6.8	36	207
8	2	2	3	404	203	6.3	1	4.7	53	143
9	2	2	4	390	271	4	1	2.4	40	77
10	2	3	3	389	203	5.7	0	5.2	0	181
11	2	3	4	406	271	3.8	0	3.3	1	128

Simulation has been executed for the equivalent of one day of activity, considering the arrival information at the Hospital of Sabadell ED, specifically during February 2012 (daily average of 397 patients, the 43.22% with level of acuity IV, the 25.35% with V), and the mix of Staff informed by such hospital (2 admission staff with senior profile, 2 triage nurses with junior profile, and 3 doctors, 1 senior and 2 juniors). Concerning the derivation percentage of patients with the lowest level of acuity, level V (the percentage of patients with level V that are sent to other health care units, and leave the ED after the triage process), six different scenarios have been considered: 0%, 20%, 40%, 60%, 80% and 100%. For each percentage four simulations have been executed, each one of them with a different seed. The data taken for each percentage of derivation is the average of the results obtained in the four executions.

The ED performance has been measured through three indicators: 1) number of patients of each acuity level (IV and V) who complete the process during the execution; 2) the percentage that treated patients represent over the number of patients arrived (calculated separately for patients IV and V); 3) and Length of Stay. The results obtained are shown in figures 7 and 8.

With 0% derivation (which means that all the patients with a level V of acuity are attended in the diagnostic and treatment zone), only 136 patients complete the process during the execution (51 of kind V and 85 of kind IV), in front of 174 (60 of kind V and 114 of kind IV) when the derivation is 100% (all the patients with a level V of acuity leave the ED after the triage process). This represents an increase of the 27.94% in the patients attended (34.12% in the case of patients IV, and 17.65% for patients V).

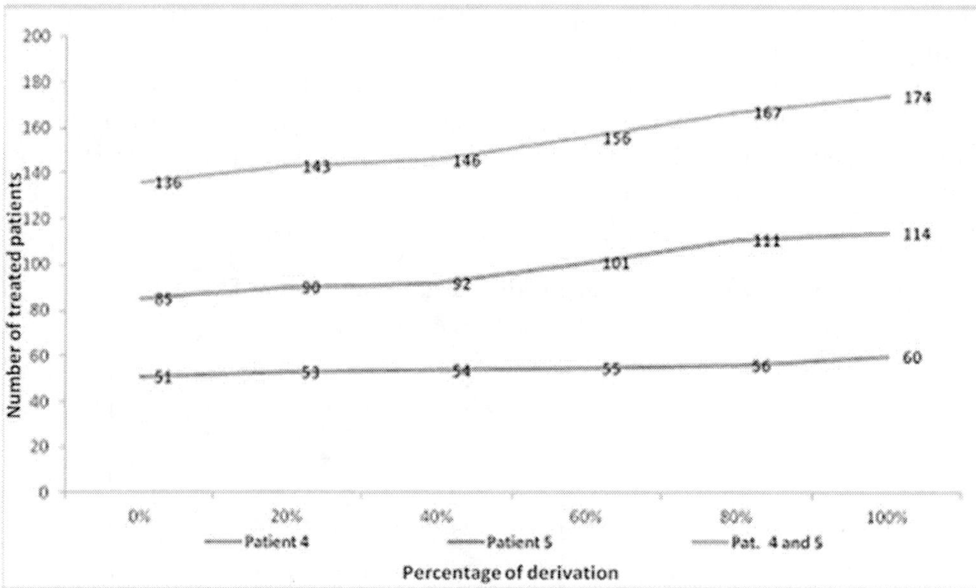

Figure 7. Number of patients that complete the ED process during the execution for different percentage of derivation (from 0% to 100%).

Considering the LoS, with 0% of derivation patients remaining in the ED for an average of 4.36 hours (4.12 hours in the case of patients of kind V and 4.36 for patients of kind IV), in front of an average 3.73 hours (3.18 in patients of kind V and 3.90 in patients of kind IV) when the derivation is 100%, which represents a reduction of 11.80% considering all patients (29.50% in the case of patients IV, and 11.79 % for patients V). In conclusion, both the number of patients treated and the LoS improve with the increase of the % of derivation.

Analyzing the occupation level of each one of the physicians, calculated as the percentage of time that a physician is attending or interacting with a patient, in relation with the total time available for attending patients, the occupation level is lower as the percentage of derivation increases (66.01% with 0% of derivation, 53.13% with 40 % and 44.71% with 100% of derivation). Taking this information into account, a new execution was carried out, now reducing the number of doctors to 2, both with junior profile, and considering a 100% derivation. With this new mix of Staff, the occupation level increases to 63.96% and the LoS of patients with a IV level of acuity is 4.21 hours, which means an increase of 18 minutes. In conclusion, despite the lower number of physicians, a very similar performance is achieved.

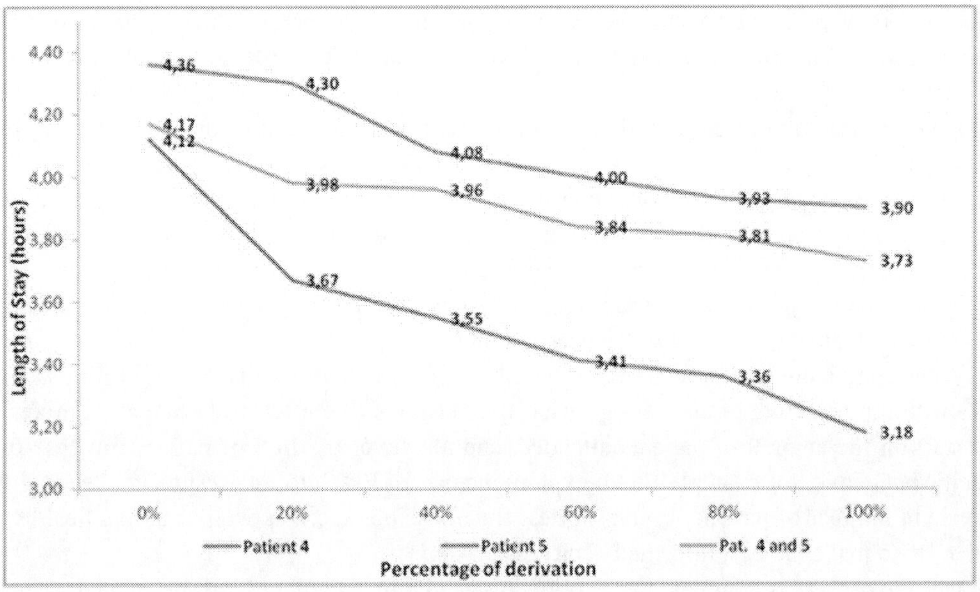

Figure 8. Average LoS (in hours) of patients that complete the ED process during the execution for different % of derivation (0% to 100%).

Therefore, and as was expected, the results obtained show that the derivation of patients that don't require urgent attention leads to a reduction in the LoS of the patients that remain in the ED. Even more than the number of physicians required in the zone in which patients are attended with level of priority IV or V is lower, which makes it possible to move some of them to other parts of the ED. It is important to take into account that these results are not conclusive because the experiments have been carried out before the tuning process that ensures the proper adjustment of the simulator with the ED of the Hospital of Sabadell.

Model Validation

The most definitive validation method is to compare the output data from the simulation with the actual data from the existing system using formal statistical analyses such as confidence intervals. This will be done following a *two-step process* described below.

The first step will be to carry out a *tuning process* between the simulation and the real system. *Configuration variables* (number of admission desks, number of triage boxes, number of attention boxes, number of the different kind of ED staff, etc) and *input data* (number and kind of patients that arrive at the ED during each one of the hours of the period of time considered) from the hospital (real system) will be introduced in the simulator, and parallel simulations will be performed changing the value of the parameters that have to be adjusted (time required by each one of the kinds of ED staff for completing each one of the processes that each kind of ED staff has to carry out) in each one of them. Results obtained (average time that each kind of patient spends in each one of the parts of the ED, number of patients treated during the period of time considered, etc,) will be compared in order to identify if the similarity between simulation and real achieves the requested level. The value of parameters that produce the best similarity level will be selected and incorporated into the simulator.

Once the first step has been completed, the second step will consist of testing its predictive power (predictive validation), using the model to predict the system's behavior, which means comparing output data of the system's behavior with the model's forecast to determine if they are similar enough. Information from the same hospital will be necessary but now referred to a different period. The information related to the configuration variables and input data will be introduced into the simulation, this time with the values of the parameters obtained in the tuning phase. Results obtained through the execution of the simulation will be compared with data from real system for identifying its similarity level and if the predictive ability of the simulator achieves the proper level. To obtain a high degree of confidence, comparisons of the model's and system's output behaviors for several different sets of experimental conditions will be carried out.

After completing the process described above, the simulator will be ready to be used as DSS with the objective of answering "what if" questions (the effects of alternative mixes of ED staff on the amount of patients attended, and also over the LoS of patients; in case of a specific pandemic, for example flu, to explore the mix of ED Staff that results in attending the maximum number of patients in the shortest time possible; etc.), as well as aiding healthcare managers to make the best informed decisions possible.

Once the process is completed, the following validation techniques will be applied:

- Internal validity: several replications (runs) of a stochastic model will be made to determine the amount of (internal) stochastic variability in the model.
- Parameter variability: or sensitivity analysis. This technique consists of changing the values of the input and internal parameters of the model to determine the effect upon the model's behavior or output, in order to check if the sensitivity of the real system is similar to the sensitivity in the model.

Before using the simulator in a different Hospital ED the process described above should be repeated (now with Hospital data) for ensuring its predictive capability in relation to this specific Hospital. The *generality* of the simulation is achieved through this process divided in two phases.

4. Optimization

4.1. Description of the Problem

The optimization problem considered in this chapter aims to find the best emergency department staff configuration (doctors (D), triage nurses (N), and admissions (A)), each of which may have two possible levels of expertise: low and high, labeled as junior and senior respectively. A junior will require more time to finish a specific process than a senior member of staff, but obviously a senior is more expensive in terms of salary than a junior. Time and salary of each type, and the amount of them -from 1 up to 3 or 4- are shown in the Table 5.

Table 5. Information related to the different types of ED Staff: salary, time required for completing the process that they have to carry out, and the minimum/maximum quantity of them

	Time (ticks)		Salary (€)		*Quantity*
	Senior (S)	Junior (J)	Senior (S)	Junior (J)	
Doctor (D)	260	350	1000	500	1 – 4
Nurse (N)	90	130	500	350	1 – 3
Admin (A)	20	35	200	150	1 – 3

This represents a combinatorial or multidimensional problem, in which each variable (in this case each ED staff member) represents one dimension, which means having three dimensions up to now, plus the patients arrival. As has been summarized in the introduction section of the chapter, our optimization problem may be expressed through the equation (1):

max / min f (x)
subject to $x \in C$

where x is the variable, *f* is a function ($f: C \rightarrow R$), *C* is the constraint set, and $\exists X_o \in C$ such that f (x_o) ≤ f (x) $\forall x \in C$ for minimization, and f (x_o) ≥ f (x) $\forall x \in C$ for its counterpart, maximization. x is discrete, global sampling from x is possible, and f (x) at any $x \in C$ cannot be evaluated exactly, and must be estimated via a simulation procedure. As it has been said, the objective is to identify the mix of the ED staff that optimize the performance of the emergency department, which due to the complexity of the system, needs to be estimated via a simulation.

Figure 9 shows such a combinatorial problem (figure 9a shows it for doctors, while figure 9b for nurses and figure 9c for admission personnel) that can take a lot of time to be solved.

This implies that specific configurations have to be simulated several times, changing the parameters to show different probabilities, for the purpose of generating a set of results from which particular effects can be concluded. In this initial approach the search for the optimum was done through an exhaustive trawl, even though it implies a lot of search time, it is guaranteed that the optimum was found.

The simulator is used as a black box, even though the more realistic the simulator is, the better results and optimizations are. The present scenario adopted for the experiments is to simulate patients moving through a simplified ED that includes four primary areas: admissions, 3 triage booths, 2 waiting rooms, and the 4 diagnosis and treatment booths. For simplicity, only four different types of active agents are considered: patients, admission staff, nurses, and doctors. ED staff members have two levels of expertise: junior or senior. In order to validate the simulator, basic patient attributes set as well as simple patient flow have been defined. Patients arrive to the ED by their own means, and wait to be attended in the admission area. Then, patients wait in the first Waiting Room (WR), WR1, until a triage nurse calls them. After the triage process patients go to a second WR2, and wait there until a free doctor call them to begin the diagnosis, and treatment, which depends on the patient's symptoms, and physical condition, as well as prescribed tests. In the end, patients are discharged from the ED. Even though realistic treatment is based on the acuity of patients, in this simulation patients follow the same path throughout the ED. Concerning to the patients' arrival, four different scenarios have been considered (84, 186, 298 and 394 patients), identified as 20, 40, 60 and 80 respectively.

DR1	DR2	DR3	DR4
DS	-	-	-
DJ	-	-	-
DS	DS	-	-
DJ	DJ	-	-
DS	DJ	-	-
DS	DS	DS	-
DJ	DJ	DJ	-
DS	DJ	DJ	-
DS	DS	DJ	-
DS	DS	DS	DS
DJ	DJ	DJ	DJ
DS	DJ	DJ	DJ
DS	DS	DJ	DJ
DS	DS	DS	DJ

(a) 14 Doctor configurations. DRi represents Diagnosis Room "i"

TR1	TR2	TR3
NJ	-	-
NS	-	-
NJ	NJ	-
NS	NS	-
NJ	NS	--
NJ	NJ	NJ
NS	NS	NS
NJ	NS	NS
NJ	NJ	NS

(b) 9 Nurse configurations. TRi represents Triaje Room "i"

A1	A2	A3
AJ	-	-
AS	-	-
AJ	AJ	-
AS	AS	-
AJ	AS	--
AJ	AJ	AJ
AS	AS	AS
AJ	AS	AS
AJ	AJ	AS

(c) 9 Admission staff configurations. Ai represents Admission Desk "i"

Figure 9. Combinations of sanitarian staff: Doctors (D), Nurses (N) and Admission (A) personnel. Two kinds of expertise: Junior (J), and Senior (S).

Even with this much reduced setting of an ED the search space is large. Such search space has 4,536 combinations to find out which is the best or optimum that minimizes a desire index, under some restrictions. These combinations result in assuming a combination of two types -junior or senior- of up to 4 doctors, 3 nurses, 3 admissions (showed in Figure 1), and 4 different scenarios of patients arrival (i.e. $14 \times 9 \times 9 \times 4$). The period simulated was 24 hours in one day, which represent 25,000 ticks for all the experiments, as well as the same random seed, and an average input of 400 patients, which is the average incoming patient that the heads of Hospital of Sabadell ED we are working with have reported.

One index was set in order to evaluate the utility of the Agent-Based ED simulator for optimizing resources. All simulations were carried out using the simulator already explained, using the BehaviorSpace tool, serially and using an IBM cluster, which has 32 compute nodes with 2 x Dual-Core Intel(R) Xeon(R) CPU 5160 running at 3.00GHz, with 12 GB of RAM, and 4MB of L2 share cache (2x2).

4.2. Experiment

The objective of this index is to minimize patients' LoS, with a budget constraint lower than or equal to 3,500 e. This cost restriction is formed by the salary of each member of healthcare staff that constitutes the corresponding configuration and which can be represented as follows: *Cost* = $\Sigma\beta_{ij} \times \lambda_{ij}$; where β=*person* and λ = *salary*, i = D, N, A and j = S, J. Thus, such an index is expressed mathematically in equation (2).

$$\begin{aligned}&\text{Minimize length of stay (LoS) } f(D, N, A)\\&\text{subject to Cost} \leq 3{,}500\end{aligned} \tag{2}$$

The results are shown in Figures 10 and 11; where the blue points are the staff configuration that satisfies the constraint, while the green and red points are the minimum for each case. Each staff configuration that got the minimum is presented in Figures 10c, 10d, 11c, and 11d, as well as the cost of the staff configuration. From Figures 10a and 10c, there are three different staff configurations that got the minimum time, but with different costs. Also, in the same Figure 10a it can be appreciated that there are many other staff configurations that are quite close to the minimum time, but with less cost.

In the other cases, where patient arrival increases, i.e. there is a higher probability; there are only few staff configurations around the minimum, or clearly only one. Not only does the patient arrival increase, but so does the minimum average patient stay time, as expected, as well as the cost of the staff configuration, and also the standard deviation of the average patient stay time. The amount of patients increases in waiting rooms, both WR1 and WR2 at times t_1, t_2, t_3, and t_4 (each time represents every 7,500 ticks of simulation), and finally the amount of unattended patients increases as well. In Table 6 all these results are shown. It is noticed when the patient arrival is high, scenario 80, patients in waiting rooms increase, i.e. the ED becomes overcrowded.

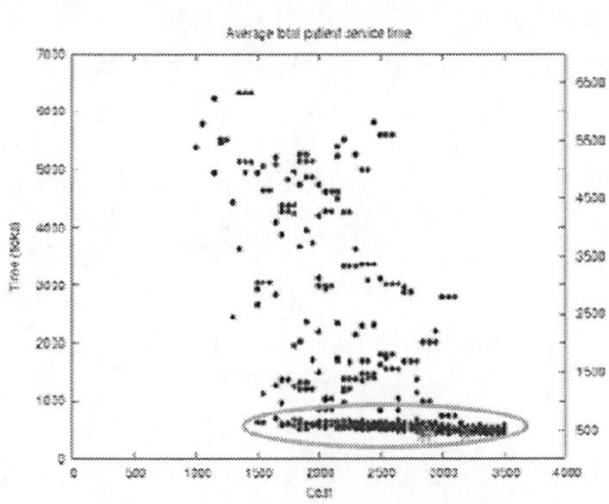

(a) Average patient LoS for arrival of 84 patients

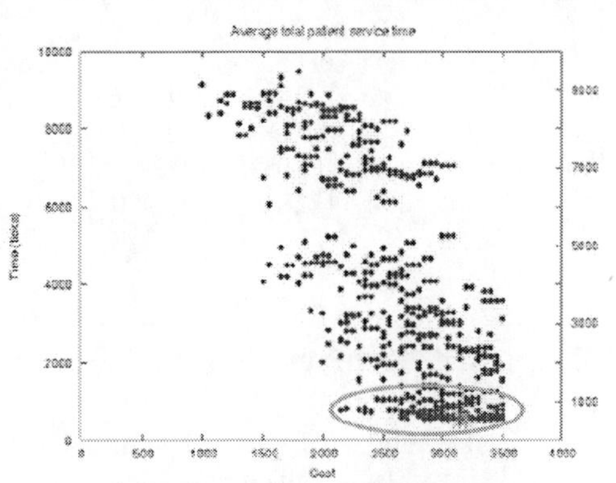

(b) Average patient LoS for arrival of 186 patients

Min	€	Time (ticks)	# staff	D	N	A
1	3,200	428	5	2 S	2 S	1 S
2	2,900	428	5	2 S	1 S	2 S
3	2,850	428	5	2 S	1 S	1 S, 1 J

(c) Staff configuration that got the average minimum time for an arrival of 84 patients. They are shown in green and red colour in Figure 10 (a)

Min	€	Time (ticks)	# staff	D	N	A
1	3,150	514	5	2 S, 1 J	1 S	1 J
2	3,200	514	7	4 J	2 S	1 S

(d) Staff configuration that got the average minimum time for an arrival of 186 patients. They are shown in green and red colour in Figure 10 (b)

Figure 10. Average patients' LoS. Red and green color points are the minimum.

It is worth pointing out that each of the plotted points were being run on the simulator as many times as points there are, i.e. for each case the optimum red and green points were obtained through an exhaustive search, which implies that all the scenarios that satisfy the restrictions should be executed on the simulator. This implies a lot of computing time, even though parametric executions, through NetLogo's Behavior Space tool were used. Therefore, when an entire and not a simplified ED is used, the number of cases that have to be tested through an exhaustive search will be extensive. In order to reduce computational cost an intelligent search, rather than exhaustive, must be used.

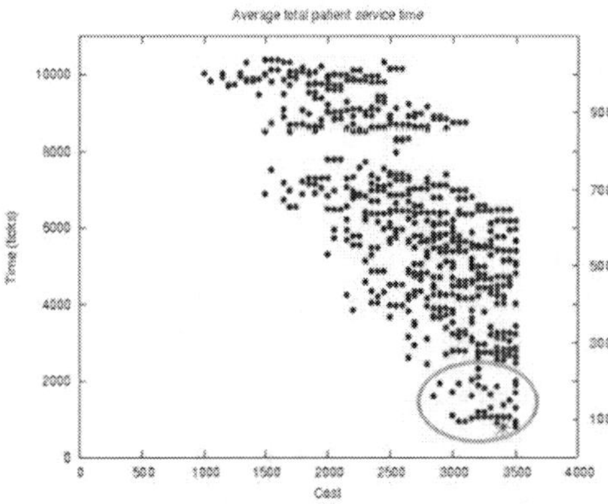

(a) Average patient LoS for arrival of 298 patients

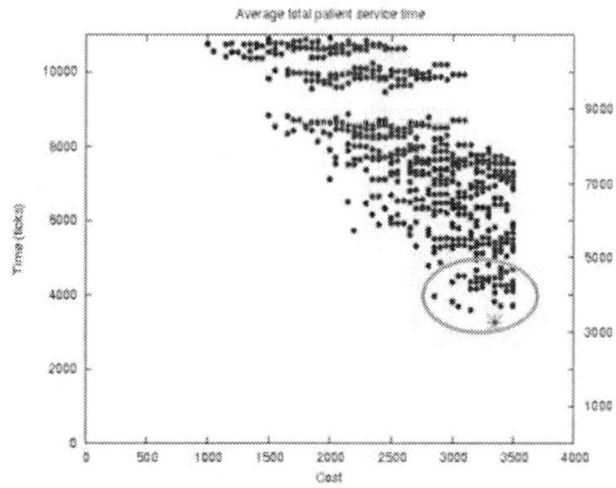

(b) Average patient LoS for arrival of 394 patients

Min	€	Time (ticks)	# staff	D	N	A
1	3,400	790	7	1 S, 3 J	2 J	1 S

(c) Staff configuration that got the average minimum time for an arrival of 298 patients. They are shown in green and red color in Figure 11 (a)

Min	€	Time (ticks)	# staff	D	N	A
1	3,350	3,266	7	1 S, 3 J	2 J	1 J

(d) Staff configuration that got the average minimum time for an arrival of 394 patients. They are shown in green and red color in Figure 11 (b)

Figure 11. Average patients' LoS. Red color points are the minimum.

Table 6. Results for the best average (minimum) for each one of the four scenarios

Scenario	Time (ticks)	σ avg. time	Cost (€)	attended patients	unattended patients	WR 1 t1, t2, t3, t4	WR 2 t1, t2, t3, t4
20	428	48 (11%)	2,850	83	1	0,0,0,0	0,0,0,0
40	514	81.5 (15.9%)	3,150	182	4	0,0,0,1	0,0,0,0
60	790	174.5 (22.1%)	3,400	290	8	1,2,0,1	3,2,4,1
80	3,266	1670.4 (51.2%)	3,350	294	100	8,19,32,43	12,25,37,51

4.3. Proposed Reduced Scheme

On the one hand, in Figures 10a and 10b there are different staff configurations that got the minimum time, but with different costs, as well as there being many other staff scenarios that are quite close to the minimum time, but with a lower cost, i.e, these other solutions are sub-optima. On the other hand, when patient arrival increases, there are only few staff configurations around the minimum, as can be seen in Figures 11a and 11b. These configurations lie near the region where the optimum is, and they are identified by red circles in Figures 10 and 11. The red circle delimits a region, which is not only where the optimum is, but also where sub-optima can be found. Some of such sub-optima configurations could be selected instead of the optimum one, because with a little less LoS they have a cheaper cost. The region is identified visually through Figures 10 and 11, and it can be summarized in Figure 12. In this figure the blue points that lie in the red parabola delimit a wide region where different configurations yield different LoS with huge difference in cost, i.e. the region is wide or there are not many points near the minimum. On the contrary, the green curve and the blue points in it represent a broad region with many points near the optimum. These other points are sub-optima and can perfectly be chosen as a solution to the problem.

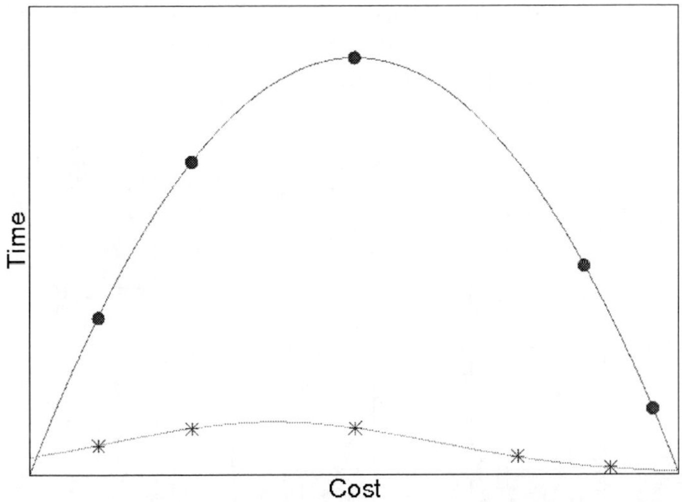

Figure 12. Two different curves where clearly exists only one optimum in the red parabola, while in the green curve there are points nearby the optimum, i.e. they are sub optima.

It was identified that when ordering simulation points (ordered by ticks or weighted time (t*) of staff configuration), before plotting it resembles the operation of a pipeline. The ordered and non-ordered data are shown in Tables 7, 8 and 9. Thus, a computationally simplified model of ED as a pipeline scheme was implemented in order to verify this assumption. The pipeline scheme proposed is described as follows: patients continue arriving at the service, when only one admission personnel, nurse and doctor are available the ED can be seen as a single queue system, whereas when more than one member of staff is available patients are attended as if they are many parallel servers. When the pipe is full, i.e. all healthcare staff (admission personnel, nurses and doctors are busy) and the previous queues of triage and diagnostic are occupied, then every certain ticks one patient being discharged from the ED (ideally, because it depends on the patient's acuity). This weighted time can be expressed as in equation 3:

$$\text{Pipeline time } t^* = \frac{1}{\sum \frac{1}{A_i}} + \frac{1}{\sum \frac{1}{N_j}} + \frac{1}{\sum \frac{1}{D_k}} \tag{3}$$

Therefore, the proposal is to use a computationally simplified model of ED which consists of pipelines, i.e. the ED can be viewed as a set of pipelines connected among them. This can be observed in Figure 13. This alternative option to find the optimum, or the region where the optimum is, can lead us to find out the optimum or sub-optimum values in a clever way, reducing the computing time and the region where the optimum lies. Once the region has been identified using this scheme, the Agent-Based ED simulator can be executed exhaustively on it. The Figure 14 shows a scattered 3D graph of Figure 10a. Using this graph nothing can be concluded about a region where the optimum and sub-optimum are, whilst the same data but previously ordered according to the first column of the Tables 7, 8 and 9, and obtained through the computationally simplified model of ED, the pipeline scheme, is shown in Figure 15. In this graph the region where the optimum and suboptimum lie is clearly identified. In such a region, an exhaustive but reduced search may be carried out in order to find out the optimum by using an intelligent way.

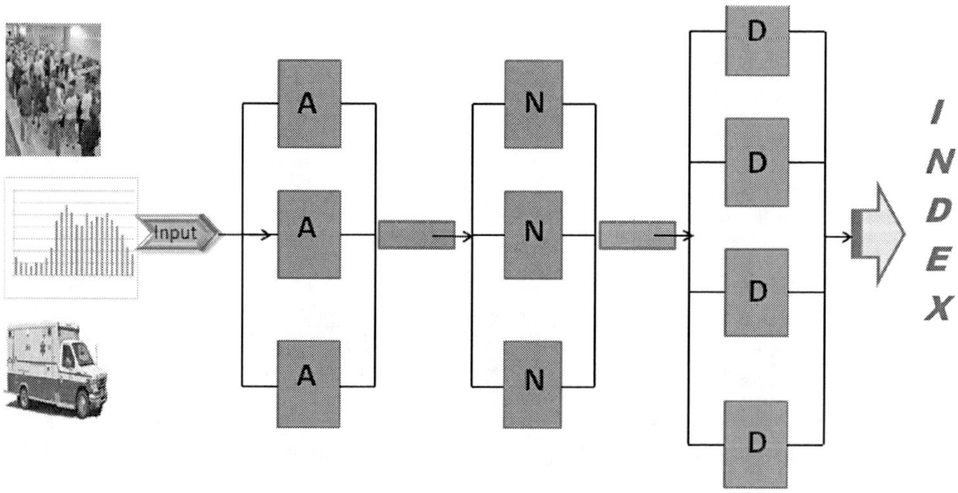

$$t^* = 1/\sum 1/A_i + 1/\sum 1/N_j + 1/\sum 1/D_k$$

Figure 13. The proposal ED computationally simplified.

Table 7. Ordering data according to working time of each staff configuration of Doctors

Ordered t*	Non ordered	DR_1	DR_2	DR_3	DR_4	€	t (ticks)	t * (ticks)
1	10	DS	DS	DS	DS	4000	260	65
2	14	DS	DS	DS	DJ	3500	350	69.5
3	13	DS	DS	DJ	DJ	3000	350	74.6
4	12	DS	DJ	DJ	DJ	2500	350	80.5
5	6	DS	DS	DS		3000	260	86.7
6	11	DJ	DJ	DJ	DJ	2000	350	87.5
7	9	DS	DS	DJ		2500	350	94.8
8	8	DS	DJ	DJ		2000	350	104.6
9	7	DJ	DJ	DJ		1500	350	116.7
10	3	DS	DS			2000	260	130
11	5	DS	DJ			1500	350	149
12	4	DJ	DJ			1000	350	175
13	1	DS				1000	260	260
14	2	DJ				500	350	350

Table 8. Ordering data according to working time of each staff configuration of Nurses

Ordered t*	Non ordered	TR_1	TR_2	TR_3	€	t (ticks)	t * (ticks)
1	6	NS	NS	NS	1500	90	30
2	9	NS	NS	NJ	1350	130	33.4
3	8	NS	NJ	NJ	1200	130	37.7
4	7	NJ	NJ	NJ	1050	130	43.3
5	3	NS	NS		1000	90	45
6	5	NS	NJ		850	130	53.2
7	4	NJ	NJ		700	130	65
8	1	NS			500	90	90
9	2	NJ			350	130	130

Table 9. Ordering data according to working time of each staff configuration of Admission personnel

Ordered t*	Non ordered	A_1	A_2	A_3	€	t (ticks)	t * (ticks)
1	6	AS	AS	AS	600	20	6.7
2	9	AS	AS	AJ	550	35	7.8
3	8	AS	AJ	AJ	500	35	9.3
4	3	AS	AS		400	20	10
5	7	AJ	AJ	AJ	450	35	11.7
6	5	AS	AJ		350	35	12.7
7	4	AJ	AJ		300	35	17.5
8	1	AS			200	20	20
9	2	AJ			150	35	35

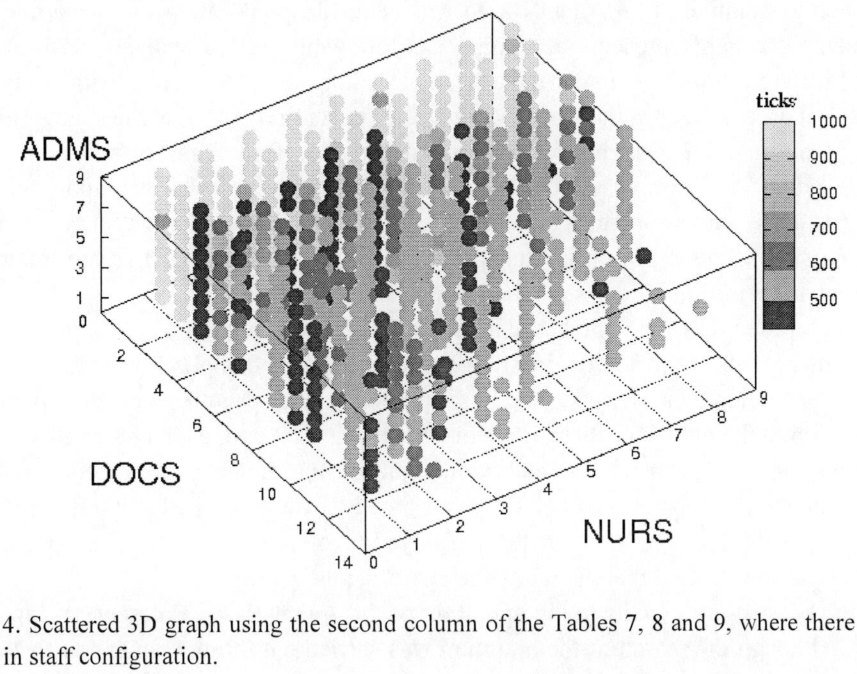

Figure 14. Scattered 3D graph using the second column of the Tables 7, 8 and 9, where there is a lack of order in staff configuration.

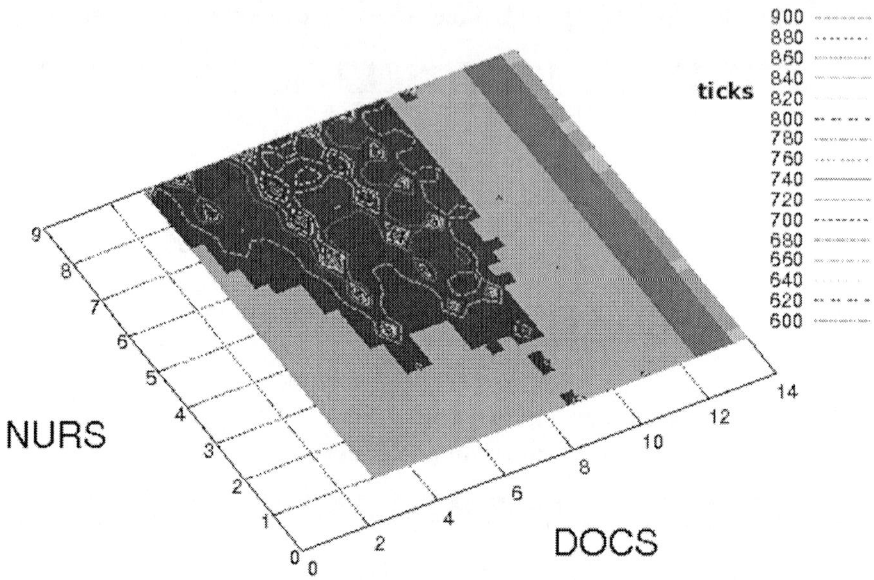

Figure 15. Isosurfaces graph using the first column of Tables 7, 8 and 9, ordered by weighted time pipeline (t*).

Conclusion

A concrete example of an Agent-Based Model for hospital emergency departments has been presented, which represents a hospital ED-following system analysis performed at a number of different hospitals, under the advice of healthcare professionals with many years' experience. It is a pure Agent-Based Model, formed entirely of the rules governing the behavior of the individual agents which populate an ED. Two kinds of agents have been identified, active and passive. Active agents represent people who act upon their own initiative (patients, admission staff, sanitarian technicians, triage and emergency nurses, and doctors), passive agents represent systems that are solely reactive, such as the loudspeaker system, the patient information system, and central diagnostic services.

The model uses state machine based agents which act and communicate within a defined environment, providing the ability to study the dynamic of complex systems without the difficulty of obtaining exhaustive system descriptions required by other modeling paradigms. The simulation of the model has been implemented using NetLogo, an agent-based simulation environment, and the verification and validation of the model have been carried out displaying the model's operational behavior graphically, examining its logic through a varied set of experiments, and finally with the participation of the ED Staff of the Hospital of Sabadell, who have verified that the model behaves like the real system.

One index was set to evaluate the operation of the Agent-Based Emergency Department simulator. Although the search of the optimum was carried out through an exhaustive search, which implies a lot of search time, the results were encouraging, since they not only showed what was expected, that the greater the amount and experience of the staff, the lower the average patient LoS is, but they also showed interesting results when standard deviation is analyzed. Simulation allows us to understand and to analyze the problem better. Nevertheless, even with this pretty small problem size the amount of combinations is large, as is the execution time. Moreover, the resources needed in order to carry out statistical sensitivity analysis for longer periods are huge, first to reproduce and then to forecast. Therefore, a better approach optimization, rather than exhaustive search, must be used. This approach consists of the use of a computationally simplified model of ED, as a pipeline scheme. The results using such an approach have not only been promising, but also computationally not as intense as the computing using exhaustive search.

In the next stage the definitive validation method will be carried out by comparing the output data from the simulation with the actual data from the existing system using formal statistical analyses. This will be done following the *two-step process* described above, by first executing several replications (runs) to determine the amount of (internal) stochastic variability in the model, and second by a sensitivity analysis. This will require applying parallelization techniques and High Performance Computing. From this point, the simulation will be ready to be used as the core component of a decision support system to aid hospital administrators make better use of resources, achieving a more efficient and improved patient care cycle. This in turn will allow better management of dynamic patient flow, either as a result of specific circumstances (pandemics, disasters, etc) or seasonal fluctuation.

In addition, the proposed alternative methodology scheme, the pipeline scheme has to be tested with a more detailed ED facility, as well as more healthcare staff and a more realistic budget. Again, due to the multidimensional nature of the problem, i.e. a large amount of individuals as well as the number of states in the state machine of each individual, different time periods, and the multi-objective optimization problem, a large amount of values should be computed; therefore, the use of High Performance Computing will be mandatory.

Acknowledgment

The research presented in this chapter has been supported by the MICINN Spain, under contract TIN2007-64974, and the MINECO (MICINN) Spain under contract TIN2011-24384.

About the Authors

M. Taboada is an associate professor in the Tomas Cerda Computer Science School (University Autonoma of Barcelona - UAB), Spain. He studied Business Administration at the University of Barcelona, Spain. From 2009 is Ph.D. student at the Computer Architecture and Operating System (CAOS) Department of the UAB. He is mainly interested in the simulation-based decision support for the operational management of Healthcare units and organizations, and specially the Hospital Emergency Departments. His email address is manel.taboada@eug.es.

E. Cabrera is a PhD student of High Performance Computing at University Autonoma of Barcelona. He received his master degree from the Computer Architecture & Operating Systems at University Autonoma of Barcelona and a B.S. degree in Computer Engineering from Universidad Nacional Autónoma de México. His re-search interests include Modeling and Simulation, Agent Based Models, Computer Architectures, Parallel Programming and Optimization. His email address is ecabrera@caos.uab.es.

E. Luque is a full professor in the Computer Architecture and Operating System (CAOS) Department at University Autonoma of Barcelona (UAB), Spain. From 1994 he has been involved, leading his group, in several re-search projects founded by the European Union (EU), the Spanish government and different industries. His major research areas are: parallel computer architecture, parallel and distributed simulation, performance prediction and efficient management of multicluster-multicore systems and fault tolerance in parallel computers and has published more than 40 papers in technical journals and more than 180 papers at international conferences.

F. Epelde (MD, Ph D) is Internal Medicine Doctor. Currently Medical Coordinator ot the Short Stay Unit in the Emergency Department in the Hospital de Sabadell. Consorci Hospitalari i Universitari Parc Taulí in Sa-badell (Spain) and professor of Emergency Medicine in the Medicine School of the UAB. e mail: fepelde@tauli.cat.

M. Luisa Iglesias (MD), currently is Director of the Emergency Department in the Hospital de Sabadell - Consorci Hospitalari i Universitari Parc Taulí in Sabadell (Spain) and professor of clinical toxicology in the Medicine School of the UAB. E-mail: miglesias@tauli.cat.

References

[1] Ahmed M. A., T. M. Alkhamis, Simulation optimization for an emergency department healthcare unit in Kuwait, *European Journal of Operational Research* 198 (3) (2009) 936 – 942.

[2] Allan, R. *"Survey of Agent Based Modelling and Simulation Tools"* Last modified October 2010. http://epubs.cclrc.ac.uk/bitstream/5601/DLTR-2010-007.pdf.

[3] Bonabeau, E. "Agent-based modeling: Methods and techniques for simulating human systems" *Proceedings of the National Academy of Sciences* 99 (2002) 7280–7287.

[4] Brailsford, C. S. 2007 "Tutorial: Advances and Challenges in Healthcare Simulation Modeling" In *Proceedings of the 2007 Winter Simulation Conference* edited by S. G. Henderson, B. Biller, M.-H. Hsieh, J. Shortle, J. D. Tew, and R. R. Barton, eds.

[5] Escudero-Marin, P., M. Pidd. "Using ABMS to simulate emergency departments" *Proceedings of the 2011 Winter Simulation Conference.*

[6] Fernández, A. Pijoan J. Ares M. Mintegi S. and Benito F (2010) *Evaluación de la escala canadiense de triaje pediátrico en un servicio de urgencias de pediatría europeo In Emergencias* 2010; 22: p.p. 355-360.

[7] Fu M. C., Feature article: Optimization for simulation: Theory vs. practice, *INFORMS Journal on Computing* 14 (3) (2002) 192–215.

[8] Günal, M. *"Simulation Modelling for Understanding Performance in Healthcare"* Department of Management Science, Lancaster University, Lancaster.

[9] Hancock W. M., P. F. Walter, The use of computer simulation to develop hospital systems, *SIGSIM Simul. Dig.* 10 (4) (1979) 28–32.

[10] Hutzschenreuter, A. K., P. A. N. Bosman, I. Blonk-Altena, J. van Aarle, and H. La Poutré."Agent-based patient admission scheduling in hospitals" *Proceedings of the 2008 international joint conference on Autonomous agents and multiagent systems.*

[11] Hutzschenreuter A. K., P. A. Bosman, H. Poutr´e, Evolutionary multiobjective optimization for dynamic hospital resource management, in: EMO '09: *Proceedings of*

the 5th International Conference on Evolutionary Multi-Criterion Optimization, Springer-Verlag, Berlin, Heidelberg, 2009, pp. 320–334.

[12] Jones S. S., R. S. Evans, An agent based simulation tool for scheduling emergency department physicians, in: AMIA Annual Symposium proceedings, *AMIA Symposium,* 2008, pp. 338–342.

[13] Jun J. B., S. H. Jacobson, J. R. Swisher, Application of discrete-event simulation in health care clinics: A survey, *Journal of the Operational Research Society* (1999) 109–123.

[14] Kanagarajah, A.K., P.A. Lindsay, A.M. Miller, and D.W. Parker "An exploration into the uses of agent-based modeling to improve quality of health care" *Proceedings of the 2006 International Conference on Complex Systems.*

[15] Katz RH, Borriello G. *Contemporary logic design.* Nueva York: Pearson Prentice Hall; 2005.

[16] Laskowski M., S. Mukhi, Agent-based simulation of emergency departments with patient diversion, in: *eHealth,* 2008, pp. 25–37.

[17] Norling, E., L. Sonenberg, and R. Rönnquist. "Enhancing multi-agent based simulation with human-like decision making strategies" in: *MABS*, 2000, pp. 214–228

[18] Paulussen, T.O., A. Zöller, A. Heinzl, L. Braubach, A. Pokahr, and W. Lamersdorf. "Patient scheduling under uncertainty" *Proceedings of the 2004 ACM symposium on Applied computing.*

[19] Persson J. A., P. Davidsson, S. J. Johansson, F. Wernstedt, Combining agent-based approaches and classical optimization techniques, in: *EUMAS*, 2005, pp. 260–269.

[20] Ruohonen T., P. Neittaanmaki, J. Teittinen, Simulation Model for Improving the Operation of the Emergency Department of Special Health Care, in: *Simulation Conference,* 2006. WSC 06. *Proceedings of the Winter*, 2006, pp. 453–458.

[21] Salmerón, J.M., L. Jiménez, and M. Sánchez. "Safety and efficacy of hospital emergency departments discharge from triage by nurses accredited to use an algorithm aid for the Spanish Triage System without physician evaluation" *Emergencias* 2011; 23: p.p. 346-355.

[22] Sargent R.G. "Verification and validation of simulation models" *Proceedings of the 2010 Winter Simulation Conference*, ed. B. Johansson, S. Jain, J. Montoya-Torres, J. Hugan, and E. Yücesan, 166-183.

[23] Saunders C. E., P. K. Makens, L. J. Leblanc, Modeling emergency department operations using advanced computer simulation systems, *Annals of Emergency Medicine* 18 (2) (1989) 134–140.

[24] Stainsby, H., M. Taboada, E. Luque. "Towards an agent-based simulation of hospital emergency departments" *IEEE Computer Society,* 2009, pp.536-539.

[25] Wang L., An agent-based simulation for workflow in emergency department, in: Systems and Information Engineering Design Symposium, 2009. *SIEDS* '09, 2009, pp. 19 –23.

[26] Weng, SJ. B. Cheng, S. Kwong, L. Wang, and C. Chang. "Simulation optimization for emergency department resources allocation" *Proceedings of the 2011 Winter Simulation Conference.*

[27] Wilensky U., Netlogo, northwestern University, *Center for Connected Learning and Computer-Based Modeling* (1999). URL http://ccl.northwestern.edu/netlogo.

In: Advances in Computational Modeling Research
Editor: Anna Belya Kora
ISBN: 978-1-62618-065-9

Chapter 2

COMPUTATIONAL MODELING IN MALE REPRODUCTION

Nicola Bernabò, Mauro Mattioli and Barbara Barboni
Department of Comparative Biomedical Sciences, University of Teramo, Teramo, Italy

ABSTRACT

In recent years we are seeing an increase in male infertility, with important negative consequences on our present and future life, such as population ageing, increase in expenses for National Health Services, worsening of sexual, personal, and social strains. Unfortunately, in about 50% of cases, it is impossible to perform an adequate diagnosis. This is likely due to the specific biology of male gametes. In fact spermatozoa gain their fertilizing ability only after they reside within the female genital tract, where important physical-chemical modifications, the "capacitation", occur. At present many aspects of this process are still unsatisfactory known. In our opinion this could be the reductionist approach adopted until now in the study of this topic. Then, in consideration that spermatozoa are constituted by a network of heterogeneous components that interact collectively and nonlinearly, characterized by a complex behaviour in which the whole system is more than the sum of its single components, for the first time we adopted a computational modeling-based approach to study sperm capacitation. We realized the biological network of molecular event occurring during this process in Human sperm. As a result, we have found that spermatozoa are an ideal candidate to this approach because their molecular composition is stable, it is possible to manage them outside the organism, and it is possible to in vitro test the achievement of capacitation. Topological analyses of network demonstrated that capacitation is characterized by a high signaling efficiency and robustness against random failure. In a further experiment we modelized boar sperm capacitation, separating the molecules depending on their subcellular compartment. The analysis of model suggested that actin cytoskeleton is not only a mechanical support but it could exert a key role in coordinating the capacitation-related biochemical events. This hypothesis was, then, successfully validated by a traditional in vitro approach. More recently, we compared the models representing sperm activation in sea urchin, Caenorhabditis elegans and Human. It was found that these organisms share a similar signalling and metabolism architecture, reasonably because of the maintenance of ancestral mechanisms, characterized by important biological features such as robustness against random failure and signalling fastness, rapidity and efficiency.

In conclusion, from computational models it is possible to infer some important information not otherwise obtainable, and leading to achieve important goals, such as the understanding of complex interactions that characterize spermatozoa biology, potentially opening new perspective in drug discovery and diagnosis and therapy of male infertility.

INTRODUCTION

In recent years a revolution of amazing importance is upsetting biology. In fact until 50 years ago Scientist from all branch of the life sciences though that the only possible way to study biological systems was to dissect them in their most elementary components. In 1966 Francis Crick stated that "the ultimate aim of the modern movement in biology is to explain all biology in terms of physics and chemistry", thus strengthening the concept that, since biological systems are composed solely of atoms and molecules, it should be possible to explain their behavior only using the physicochemical properties of their individual components, down to the atomic level. From a philosophical point of view, this approach could be defined as reductionist. Reductionists think that the information on isolated atoms and molecules could be sufficient to explain the behavior of the systems they made up. Now it is clear that, as Marc H. V. Van Regenmortel, elegantly claims "their situation is similar to that of an art student asking about the significance of Michelangelo's David and being told that it is just a piece of marble hewn into a statue in 1504. This is certainly true, but it evades pertinent questions about the anatomy of the statue, its creation at the beginning of the Florentine Renaissance, its significance in European art history, or even the scars on its left arm that were plastered after it was broken in three places during the anti-Medici revolt of 1527. In an analogous way, the biology, development, physiology, behavior or fate of a human being cannot be adequately explained along reductionist lines that consider only chemical composition" (Van Regenmortel, 2004). In other words the progress of biology leads the scientific community to conceive that the reductionism is unable to satisfactory explain some phenomena, such as the function of human brain in term of intelligence or imagination, the relationship among genes, proteins and metabolism, the function of immune systems, the biology of stem-cells, the pathology of cancer, the human and animal reproduction. In all these cases the systems are, in fact, characterized by a high degree of complexity: they are composed by a large number of element interacting each other in a nonlinear way. This condition has some enormously important consequences:

- *it is impossible to predict how the system will evolve after a sufficiently long time interval.* The evolution of complex systems, such as the weather, the stock market, the cell biochemical machinery, is difficult to predict, because every finished model needs to remove some information about the initial conditions, and these errors tend to increase as the simulation proceeds over time. At the limit, the residual error in the simulation could be of the same order of magnitude than the result itself. In this case, in substance, the predictions from the simulation are no longer reliable;
- *the butterfly effect*, i.e. the sensitive dependence on initial conditions, where a small change at one place in a deterministic nonlinear system can result in large differences to a later state. In 1961, Edward Lorenz was using a numerical computer model to rerun a weather prediction, and entered the decimal .506 instead of entering the full

.506127. The result of calculus was completely different, thus suggesting the strong dependence of system from its initial conditions (the so called "butterfly effect"). Now this effect is well known in different branch of science, such as meteorology, sociology, economy and biology;

- *changing the scale by which the system is observed new and unexpected proprieties will emerge*. In fact new and unpredicted system propriety can appear when a number of simple entities operate in an environment, forming more complex behaviours as a collective. It is, for instance, the case of neuron networks or ant colonies. Each neuron or each ant *per se* obeys to a few simple rules, while a large number of them, reciprocally interacting, is able to display a complex and sophisticated behaviour.

The combination of these effects could explain the interpretative failure of reductionism in front of the problems aroused by high throughput disciplines, now exploding in modern biology. Today are available technologies able to analyze and to describe large sets of data, such as 2D electrophoresis, DNA microarrays, protein chips, etc ... The data obtained from these methodologies need to be examined collectively to be correctly interpreted and managed. Then a reductionist approach is not sufficient to describe the system as a whole: it is indispensable to change our point of view, adopting a more holistic perspective which will take in account not only the molecules involved in biological processes but also their complex interactions. Fortunately, the progress in biological techniques was paralleled by the development and diffusion of computational resources devoted to the data modelization. In fact, the computer has become an indispensable tool for modern biologists, posing the basis for "dry" biology (in opposition of "wet" biology, which requires the use of traditional *in vivo* or *in vitro* experimental set-ups).

COMPLEXITY IN MALE REPRODUCTION

In this context virtually all the branch of biology take great advantages by the adoption of *in silico* resources and especially biology of reproduction which is, by definition, characterized by a high degree of complexity (Bernabò et al., 2012). In particular, male gametes display a peculiar and fascinating differentiative behavior either in the organism where they are produced and in female environment.

As first they undergo a complex process of differentiation within the male genital tract. The initial stages occur in the testes, within seminiferous tubules, where the stem cells, the spermatogonia, adjacent to the inner tubule wall divide in a centripetal direction producing immature sperm. Altogether this process is the result of different successive stages:

1) *spermatocytogenesis:* this stage leads the production of haploid spermatocytes. Diploid spermatogonia, located in the basal compartment of seminiferous tubules, undergo a mitosis to produce diploid cells, the so called primary spermatocytes. Each primary spermatocyte, procedes into the adluminal compartment of the seminiferous tubules, duplicates its DNA, and subsequently divides by meiois I to produce two haploid secondary spermatocytes. Cell divisions are incomplete and, as a consequence, the cells remain connected to one another by bridges of cytoplasm to allow synchronous development (testicular waves). In this step the random inclusion

of parental chromosomes and the chromosomal crossover takes place, then conferring to each developing spermatozoon its specific genetic pattern. Chromosomal abnormalities due to the incorrect segregation of chromosomes can give rise, in humans, to important pathological events, such as Down Syndrome, Klinefelter's Syndrome, and spontaneous abortion.

2) *spermatidogenesis*: at this stage secondary spermatocytes rapidly enter meiosis II and divide to produce haploid spermatids.
3) *spermiogenesis:* the spermatids begin to morphologically differentiate. The Golgi apparatus surrounds the nucleus, giving rise to the the acrosome. One of the centrioles elongates to give rise to the tail of the sperm, the thickened mid-piece develops, and the mitochondria acquire their definitive localization. In addition the DNA becomes highly condensed. In fact histones are replaced firstly by transition proteins, then by protamines, which confer to the DNA its highly packaged structure, thanks their arginine residues.

Then, the spermatozoa are transported to the epididymis where they complete their biochemical maturation and are stored until the ejaculation. At this stage, male gametes are still unable to fertilize a mature oocyte. To gain the full fertilizing ability they must complete a second maturation process that, in organisms characterized by internal fertilization as the mammals, takes place within the female genital tract and requires from hours to days. This event, called "the capacitation", has been discovered in early 50th by Austin (Austin, 1951) and Chang (Chang, 1951), independently and 60 years later is still under the attention of Researchers because its important implication both for basic (cell biology, development biology, biochemistry) and applied science (endocrinology, andrology, study of infertility).

The biochemical modifications which happen during capacitation involve all the structure of sperm cells, from the membrane to the cytosol and the cytoskeleton. In particular, the membranes markedly change their architecture. They are highly dynamical asymmetrical structures, either in longitudinal (membrane domains and sub-domains) (Bruckbauer et al., 2010; Gadella et al., 2008; Kotwicka et al. 2011; Venditti and Bean, 2009) and transversal (inner and outer leaflet) direction (Gadella and Harrison 2000, 2002; Harrison and Gadella 2005). In the head plasma membrane (PM) it is possible to recognize different domains characterized by specific chemical- physical and functional proprieties: the apical ridge area, the pre-equatorial area, the equatorial area and the post-equatorial area. The apical ridge contains specific zona binding proteins (O'Rand & Fischer, 1987). The pre-equatorial surface is the site where the fusion between PM and outer acrosome membrane (OAM) takes place during AR. The equatorial surface area is characterized by the presence of a hairpin like structure and it is involved in the fusion of sperm and oocyte membranes at the moment of fertilization (for review see Yanagimachi, 1994 and Gadella et al., 2008). Each domain, in turn, is a puzzle of specialized areas, known as microdomains. These microdomains are isolated, by using 0.1% Triton X-100 or similar detergent at 4°C in a discontinuous density gradient, as detergent resistant membranes (DRMs). They are small lipid ordered portions of membrane containing large amounts of cholesterol, sphingomyelin, gangliosides, phospholipids with saturated long-chain acyl chains, and proteins such as GPI anchored proteins, caveolin and flotillin. During capacitation DRMs organization undergoes evident modifications, with important consequence on the function of several proteins involved in signal transduction, such as CB1R and TRPV1, (Botto et al., 2010) and in membrane fusion (Asano et al., 2010) (see Brewis and Gadella 2010 for review).

Membrane remodeling involves also the segregation of specific classes of lipids in inner or outer leaflet of plasma membrane (Gadella et al., 2008; Hickey and Buhr 2011). In fact, the aminophospholipids phosphatidylserine (PS) and phosphatidylethanolamine (PE) are mainly concentrated in the inner leaflet, while choline phospholipids sphingomyelin (SM) and phosphatidylcholine (PC) in the outer one. This asymmetry is established and maintained by the action of several translocating enzymes:

- *aminophospholipid translocase* (also known as flippase), is responsible for transfer of PS and PE from the outer to the inner lipid leaflet;
- '*floppase*' transfers phospholipids from inner to outer leaflet;
- *scramblase* acts as a bi-directional carrier with little specificity, simply moving all four phospholipid species in both directions (inward and outward) across the membrane lipid bilayer, thus reducing phospholipid asymmetry ('lipid scrambling').

All these enzymes concur in modulating the ability of outer acrosome membrane (OAM) and plasma membrane (PM) to fuse each other, the so called fusogenicity, in order to make possible the exocytosis of acrosome vescicle content. It is of vital importance that the spermatozoa reach the right level of fusogenicity at the right moment: to correctly undergo acrosome reaction (AR) they must be stable enough to maintain the integrity of their biochemical machinery, but so unstable to be ready to fuse when the physiological stimulus, the ZP proteins, is met. Thus, a complex chain of signaling events is devoted to control the achievement of sperm membrane fusogenicity: immediately after ejaculation sperm cells are immersed in the seminal plasma, which contains a high concentration of stabilizing molecules (such as endocannabinoids) and the phospholipids scrambling is blocked. Whit the removal of seminal plasma and with the contact with female genital tract secretion, the concentration of endocannabionoids gradually decreases and that of bicarbonate increases (Barboni et al. 2011). As a consequence soluble adenylate cyclase (sAC) is activated in the sperm head and cAMP/PKA pathway became effective stimulating phospholipids scrambling in the apical plasma membrane of the sperm head. This process enable the extracellular protein to extract cholesterol, thus the membranes can complete its reorganization lipids architecture, which is the necessary prerequisite for AR (Gadella and Harrison 2000).

The cytoscheleton has a central role in conferring to the spermatozoa their tridimensional asset and it acts as a controller of membrane dynamics. In fact, in sperm cells the cytosol is virtually absent as a consequence the PM is in direct contact with the underlying cytoskeleton. It is believed that actin cytoskeleton may have a structural role during capacitation and AR: the increase in membranes fusogenicity is paralleled by the actin polymerization, to form a diaphragm between PM and OAM, thus avoiding their premature fusion. Once the capacitation is completely achieved and the physiological stimulus (ZP proteins) is met, the ZP-induced fast calcium peak causes the immediate actin depolymerisation, thus removing this steric obstacle to the membrane fusion and, ultimately, AR can happens (Breitbart et al., 2005; Breitbart and Etkovitz, 2011).

The spermatozoa, in addition, undergo a series of biochemical modifications leading the onset of a new pattern of motility. In fact, in epididymis they are completely immotile or weakly motile, while immediately after ejaculation, they begin to swim with a species-specific pattern. In addition, once exposed to capacitating agents, male gametes gradually acquire a new kind of motion, the so called "hyperactivated motility". In human, the flagella of hyperactivated spermatozoa beat less symmetrical than flagella of ejaculated ones and, as a result, they tend to vigorously swim in circles. It has been proposed that the acquisition of this motility pattern could allow to the male gametes to penetrate the oviductal mucus, the cumuls-oocyte complex extracellular matrix and, finally, the ZP (Chang and Suarez, 2010; Suarez 2008). From a biochemical point of view, it is known that acquisition of hyperactivated motility is due to the activation of a complex pathway, not yet completely discovered. It has been proposed that the interaction with activating molecules contained in female environment and the removal of inhibiting factors of male origin could promote the increase in intracellaular concentration of second messengers, in particular of Ca^{2+} and cAMP. Specifically, the phospholipase C (PLC) is activated through a heterotrimeric G protein (Gq/11)-coupled receptor (R1) and produces IP3. Binding of IP3 to its receptors (IP3R) causes an increase in cytoplasmic Ca^{2+}. The activation of membrane-associated adenylyl cyclase (AC) through high cytoplasmic Ca^{2+}, G proteins and membrane potential increases intracellular cAMP. In parallel the bicarbonate may also cause an increase in cAMP by activating the soluble form of adenylyl cyclase (sAC) directly. The increase in cAMP concentration activates cyclic nucleotide-gated channels (CNG) thus promoting the Ca^{2+} influx. In addition increased cAMP activates protein kinase A (PKA) to phosphorylate axonemal or fibrous sheath proteins and results in flagellar beating. It is thought that high cytoplasmic Ca^{2+} and Ca^{2+}–calmodulin complex are responsible for asymmetrical bending of flagella that is characteristic of hyperactivation (see Suarez 2008 for review).

Computational Modeling of Spermatozoa

The approach used until now to study the capacitation has been characterized by a top-down strategy. The whole problem (how the spermatozoa gain their fertilizing ability) has been dissected in sub-problems (the different signalling pathways, the transduction of molecular messages), and in sub-sub-problems (the molecules involved, their function, their modifications), and so on, giving rise to a myriad of data concerning hundred or thousand of molecules. By this way, in one hand, our knowledge on molecular determinants of capacitation is amazingly growth in recent years but, in the other one, the level of understanding of this phenomenon is still largely unsatisfactory. The proof is that:

- *at present it is impossible to a priori identify a capacitated spermatozoon.* The currently available tests in andrology are all *a posteriori* evaluations, with important practical limitations. The chlortetracycline (CTC), staining test is based on a UV fluorescent antibiotic, that yields different patterns of distribution on the sperm surface depending on the time and conditions of the spermatozoa incubation. These different patterns have been empirically correlated with the capacitation status of the sperm cells in several species including Human (Perry et al., 1995), mice (Fraser and Herod, 1990), bull (Fraser et al., 1995), boar (Mattioli et al., 1996), etc ... Unfortunately the mechanism by which CTC yields the different patterns, as well the biochemical events involved in capacitation that give rise to these patterns, are completely unknown. It has been proposed that they are related with the changes in the distribution of Ca^{2+}-CTC complexes bound to phospholipids in the plasma membrane that from during incubation under capacitating conditions. As a consequence, any compound that interferes whit the achievement of capacitation, without affecting these complexes, could determine the onset of a CTC staining pattern characteristic of capacitated spermatozoa in the absence of fertilizing ability. It is the case, for instance, of spermatozoa treated with cytochalasin D, a specific actin polymerization inhibitor, that display a capacitation-related pattern of fluorescence, and that are unable to undergo AR, when tested by induction of acrosome reaction by homologous ZP (Bernabò et al., 2011). This finding suggests the idea that CTC could changes as a consequence of the membrane lipid remodeling occurring during capacitation, rather than depending on the real capacitation status of male gametes. Alternatively the capacitation could be monitored by the ability of more or less physiologically relevant agent, such as the calcium ionophore A23187, ionomycin or progesterone, to induce AR. In this case, it is impossible to assume that the compounds that are able to stimulate or inhibit AR, do so by stimulating or inhibiting capacitation. In other words, acrosome reactions could occur in uncapacitated sperm when the cells are treated with compounds that bypass capacitation. The zona-free hamster oocyte penetration test is aimed to provides information on the ability of human male gametes to fuse with hamster zona-free egg. Unfortunately, this test may result in two types of criticism: 1) it differs from the physiological situation in that the ZP is present; 2) the results of conventional hamster oocyte test depends on the occurrence of spontaneous AR in spermatozoa incubated for prolonged periods in vitro. Since this procedure may involve different mechanisms and is less efficient than the biological process, false-negative results have frequently been recorded (WHO, 1996). The most reliable way to assess capacitation is to perform an *in vitro* fertilization assay (IVF). In this case also, it cannot be assumed that if a specific compound or incubation condition inhibits or decreases the outcome of IVF, it really inhibits capacitation, since fertilization is a multistep process whose results depend from several variables of male and female origin. In addition, the IVF protocols often require operative condition, such sperm egg ratio, spermatozoa concentration, coincubation time, far from physiological ones.

- *in several cases, it is impossible to perform an adequate diagnosis after seminal and clinical investigations*, as it happens in case of unexplained infertility of male origin. Recent data report that infertility affects about 7% of all men and primary testicular failure accounts for about 75% of all male factor infertility. Its etiology is still unknown in about 50% of cases, in which the only possible diagnosis is "idiopathic infertility" (Krausz, 2011).

Recently, for the first time, our group has started to use a different bottom-up approach (Bernabò et al., 2010a). We think that the data available are enough to describe a relevant part of molecular events that take place during capacitation, while what we do not know are the proprieties that emerge from their complex interactions. Thus we tried to realize a computational model that will take in account not only the molecules involved in this process but also their interactions, and that describes the whole system (the spermatozoa signaling process) allowing the observation of emerging proprieties (the fertilizing ability). This last point has, in our opinion, a pivotal importance. In fact it is impossible to infer a propriety of whole the cell, looking only at molecular level. In addition, for many reasons sperm cells are an ideal candidate for this kind of analysis:

- they are virtually transcriptionally silent, thus their protein composition is stable. As a consequence it does not exist the problem to modelize the continuous modification in cellular protein content and in molecular interactions due to the dynamical regulation of genes expression and protein transcription, typical of other kind of cells;
- it is possible to evaluate the completion of capacitation by IVF trials (in animal models);
- finally the spermatozoa are able to exert their function outside the organism in which they are produced. As a consequence they are capable of independent life (unlike the other cells) and it is possible to *in vitro* manage them in laboratory without loss of the cell function.

As modeling strategy, we adopted the biological networks formalism, starting from the consideration that it is possible to represent a series of elements interacting each other as a mathematical object called "a graph". To this aim, all the molecules involved in sperm signal transduction and energy metabolism have been inserted in a database with the specification of the interaction they carry out. As a reference we used the data concerning human spermatozoa contained in PubMed database, published in the 10 years. When the data where lacking or to fill incomplete pathways the data from other mammals, such as mouse, horse, pig, bull, etc.. were used, only if confirmed by a large consensus. The freely available and diffusible small molecules such as H_2O, CO_2, P_i, H^+,O_2 were omitted, when not necessary. More in detail the fields of the database were:

Source molecule: represents the molecule source of interaction.
Interaction: represents the nature of interaction (activation, inhibition,...).
Target molecule: represents the molecule target of the interaction.
Biological function: represents the functional meaning or the contest of interaction (glycolysis, lipid remodeling, oxidative phosphorylation,…).
Species: represents the species in which the interaction was described.
Reference: represents the bibliographic source of information.
Notes: represents all the notation such as the presence of synonyms or the intracellular location, if relevant, or the explanation of complex cellular events.

The networks have been created by using the software Cytoscape 2.6.3 (http://www.cytoscape.org) and have been spatially represented using the Cytoscape Spring-embedded Layout. This program "is based on a force-directed paradigm. Network nodes are treated like physical objects that repel each other, such as electrons. The connections between nodes are treated like metal springs attached to the pair of nodes. These springs repel or attract their end points according to a force function. The layout algorithm sets the positions of the nodes in a way that minimizes the sum of forces in the network" (Cytoscape User Manual). The size of each node was proportional to the number of its connection and the node color gradient was dependent from its closeness centrality. This parameter is computed as:

$C_c(n) = 1/avg(L(n,m))$, where L (n,m) is the length of the shortest path between two nodes *n* and *m*. The closeness centrality of each node ranges from 0 to 1 and it is a measure of how fast information spreads from a given node to the others nodes (Figure 1).

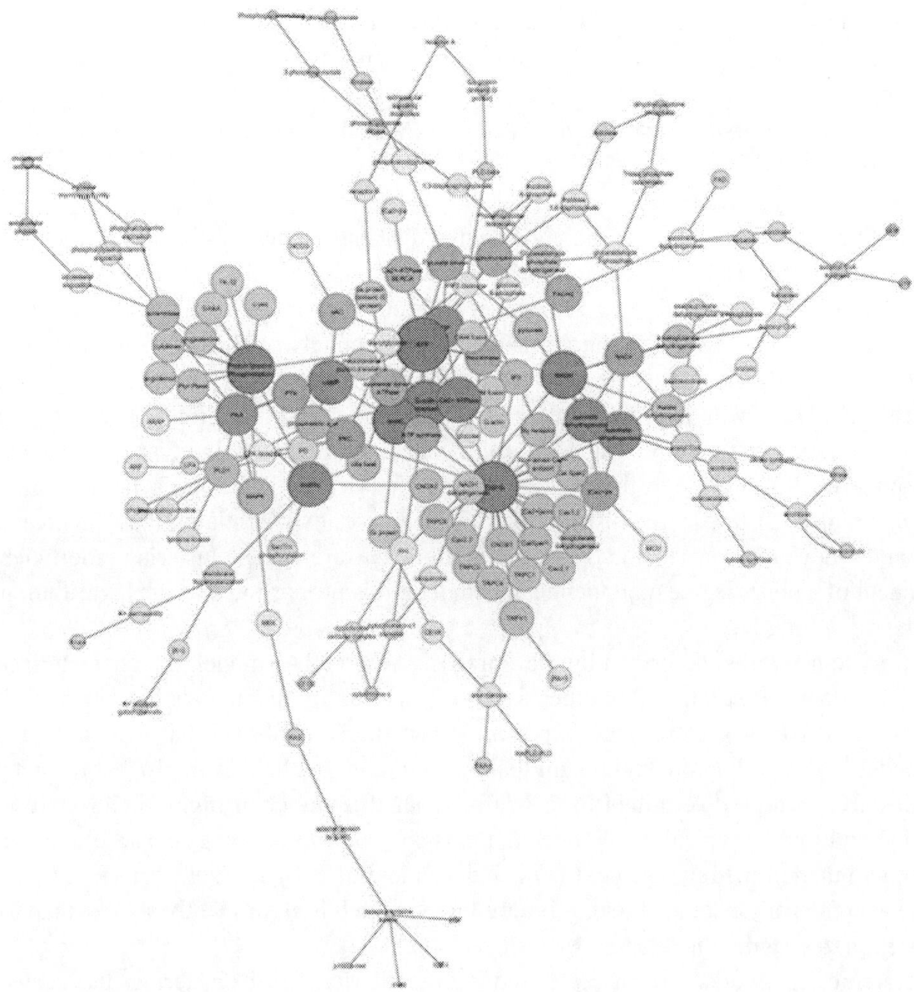

The node size is proportional to the connection number and the node color gradient is dependent from the closeness centrality. This parameter is computed as: $C_c(n) = 1/avg(L(n,m))$, were L (n,m) is the length of the shortest path between two nodes n and m. The closeness centrality of each node ranges from 0 (light gray) to 1 (dark gray) and it is a measure of how fast the information spreads from a given node to the other nodes. The spatial network arrangement is obtained by using the Cytoscape Spring-embedded Layout (see the text for explanation).

Figure 1. Diagram showing the structure of the Human capacitation network.

The statistical and topological analyses of networks were carried out considering the networks as directed by the Cytoscape plugin Network Analyzer (http://med.bioinf.mpi-inf.mpg.de/netanalyzer/help/2.6.1/index.html). In particular were assessed the following topological parameters:

- *the number of nodes*: i.e. the total number of molecules involved;
- *the number of edges*: i.e. the total number of interaction found:
- *the clustering coefficient*: that is calculated as $CI = 2nI/k(k-1)$,where nI is the number of links connecting the kI neighbours of node I to each other. It is a measure of how the nodes tend to form clusters;
- *the network diameter*: i.e. the largest distance between two nodes;
- *the averaged number of neighbours*: i.e. the mean number of connection of each node;
- *the characteristic path length*: i.e. the expected distance between two connected nodes.

In addition, for each network, the node degree distribution law has been checked, and the exponent (γ) and the result of correlation test (r) have been reported.

As it is known, it possible to recognize different typologies of networks architecture (Barabási and Oltvai, 2004). The simpler is that of random networks, described by the Erdös–Rényi (ER) model. In this case the network has N nodes connected with probability p, then giving rise to a graph with approximately $pN(N-1)/2$ randomly placed links. The node degree follows a Poisson distribution, which indicates that most of nodes have approximately the same number of links, close to the average degree, that define the scale of the network. The tail (high k region) of the degree distribution $P(k)$ decreases exponentially, thus the nodes that significantly deviate from the average are extremely rare. The clustering coefficient is independent of a node degree, and the mean path length is proportional to the logarithm of the network size, $l \sim \log N$.

Scale-free networks, described by the Barabási–Albert (BA) model, are characterized by a power-law degree distribution. In other words the probability that a node has k links follows $P(k) \sim k^{-\gamma}$, where γ is the degree exponent. Then, the probability that a node is highly connected is statistically more significant than in a random graph. This imply that the network properties are strongly determined by a relatively small number of highly connected nodes (the hubs) and that a "typical" node does not exists (scale free topology). These network do not have an inherent modularity, i.e. $C(k)$ is independent of k. In addition scale-free networks have the average path length following $l \sim \log \log N$, which is significantly shorter than log N that characterizes random networks.

Hierarchical networks are characterized by modularity, local clustering and scale-free topology. As a consequence their most important feature s the scaling of the clustering coefficient, and the hierarchical architecture implies that sparsely connected nodes are part of highly clustered areas, with communication between the different highly clustered neighborhoods maintained by a few hubs.

From the results of the analyses carried out on the network representing sperm capacitation it was evident that it could be considered a scale free network (see Table 1and Table 2).

In particular the value of clustering coefficient was very low, thus strengthening the concept that the process of acquisition of fertilizing ability by male gametes is constituted by a few signalling pathways in which the molecular message are carried, directionally, from the beginning to the end of the chain (this is also the reason why the averaged number of neighbours in about 2: one is the input and the other is the output of message). In this context, reasonably, the presence of loop or clusters could to interfere and to slow the propagation of signals.

Table 1. Main topological parameters of capacitation and AR networks

	capacitation
N° nodes	146
N° edges	197
Clustering coefficient	0.029
Diameter	20
Averaged n° neighbours	2.667
Char. path length	6.606

The number of nodes represent the total number of molecules involved, the number of edges represents the total number of interaction found, the clustering coefficient is calculated as CI = 2nI/k(k–1),where nI is the number of links connecting the kI neighbours of node I to each other, the network diameter is the largest distance between two nodes, the Averaged n° neighbours represent the mean number of connection of each node, the Char. path length gives the expected distance between two connected nodes.

Table 2. Result of power law fitting of IN and OUT capacitation network

	in	out
r	0.992	0.997
R^2	0.897	0.837
γ	-1.547	-2.046

(γ) represents the exponent of node degree law
r and R2 are the result of correlation and fitting tests.

The values of characteristic path length, of about 6.6 may have two important implications on the architecture of capacitation. As first, if any molecule can interact with any other in a small number of passages, the loss of information due to the signal decrease is minimized and the signal efficiency is maximized. In addition, any local adaptation in signalling system could reach the whole network in a very short time, thus increasing the cellular system responsiveness to the intracellular and extracellular stimuli.

The scale free topology, in addition, is characterized by an evident non-homogeneity in node connectivity. Thus, we have examine the different classes of nodes. As first we identified the hubs of system. As it is evident from Table 3, the most connected nodes are $[Ca^{2+}]_i$, ATP-ADP system, protein kynase A (PKA), Tyr phosphorylation and phospholipase D1 (PLD1).

It is known that capacitation to occur requires the increase in the $[Ca^{2+}]_i$ (Breitbart, 2002 ; Florman et al., 2008). At the present several molecular system have been proposed to be involved in this event. As clearance mechanisms the plasma membrane Ca^{2+}-ATPase exports a cytoplasmic Ca^{2+} ion and imports one or two extracellular protons at the expense of ATP. When $[Ca^{2+}]_i$ is elevated, the plasma membrane Na^{+}-Ca^{2+} exchanger operates in forward mode exporting an intracellular Ca^{2+} ion and importing approximately three Na^{+} ions at the expense of the Na^{+} gradient (Fraser et al., 1993; Blaustein and Lederer, 1999). Sarcoplasmic-endoplasmic reticulum Ca^{2+}-ATPase pumps and the mitochondrial Ca^{2+} uniporter (Wennemuth et al., 2003) operate sequestering the ion in acrosome or in mitochondria, respectively. In addition during the acquisition of fertilizing ability in sperm membrane is operative a large number of calcium channels, such as:

Table 3. Most connected nodes (the hubs) of human capacitation network

Node	Number of links
$[Ca^{2+}]_i$	25
ATP	14
Tyr phosphorylation	13
PKA	9
ADP	8
PLD1	8

- *CatSper channels*: on sperm membrane are present four pore-forming CatSper channel proteins, CatSper1–4. They contain six transmembrane-spanning (TM) domains, their the overall sequences are similar to those of CaVs and they have the T/S-x-E/D-x-W signature sequence in the ion selectivity filter region.The intracellular NH2 terminus of CatSper1 is rich in histidines (49 out of 250 amino acids in the mice), which may be related to its pH_i sensitivity. The S4 segments of all CatSpers channels have charged residues (R/K) at every third position, a signature of voltage-gated ion channels (Ren and Xia, 2010).
- *voltage-dependent calcium channels (VDCCs)*: they are a group of voltage-gated cationic channels found in spermatozoa and several others excitable cells (i.e. muscle, glial cells, neurons, etc.) with a permeability to calcium about 1000-fold greater than to sodium, under normal physiological conditions. When membranes are in the condition of resting membrane potential, VDCCs are normally closed. When membrane depolarizes, as in spermatozoa it happens during AR, they open, thus allowing $Ca2^{+}$ entry into the cell (Darszon et al., 2005).
- *transient receptor potential channels (TRP channels)*: they are about 28 channels that share some structural similarity to each other. TRP are divided into two groups: group 1 includes, TRPC ("C" for canonical), TRPV ("V" for vanilloid), TRPM ("M" for melastatin), TRPN and TRPA. In group 2 there are TRPP ("P" for polycystic) and TRPML ("ML" for mucolipin). In spermatozoa TRPV1 in particular have been found to have an important role concurring in modulation of the acquisition of fertilizing ability and the onset of AR (Bernabò et al., 2010b).
- *cyclic nucleotide-gated ion channels (CNG channels)*: they are ion channels that function in response to the binding of cyclic nucleotides cGMP and cAMP and either of a depolarization or a hyperpolarisation event (Navarro et al., 2008).

The ATP-ADP system represent the main energetic source of spermatozoa. In fact in male gametes, the production of metabolic energy is fuelled by the glycolysis exclusively, by mitochondrial oxidative phosphorylation exclusively, or by a combination of both the pathways (Storey, 2008).

The PKA is involved in several biochemical events, acting as a modulator of different pathways. In particular the HCO_3^- and Ca^{2+} are transported by a Na^+/HCO_3^- cotransporter (NBC) and a sperm-specific Ca^{2+} channel (CatSper) and promote, via soluble adenylyl cyclase, the PKA activity. This enzyme is known to be involved in late events that lead the spermatozoa to acquire hyperactivated motility and in the increase in tyrosine phosphorylation, dependently on the presence of cholesterol acceptors in the capacitation medium (Breitbart et al., 2006; Visconti, 2009).

This last event plays a key role during the capacitation in several species such as human, mice, cattle, pigs, hamsters and cats (Barbonetti et al., 2004; Visconti et al., 1995; Urner and Sakkas, 2003). The modification of tyrosine phosphorylation pattern is thought to be a necessary prerequisite for the completion of capacitation and has been demonstrated to change both in flagellum and in the sperm head, but at present the specific molecular target of protein tyrosine phosphorylation are largely unknown (Urner and Sakkas, 2003).

PLD1 is involved in control of actin polymerization, which is one of the most important events occurring during capacitation. More in particular MAP-kinase, tyrosine kinase, and ADP-ribosylation factor lead PLD activation, thus increasing the phosphatidyl-choline hydrolysis to produce phosphatidic acid, which stimulatesthe polymerization of G-actin to form F-actin (Gomez-Cambronero and Keire, 1998; Cohen et al., 2004; Breitbart et al., 2005).

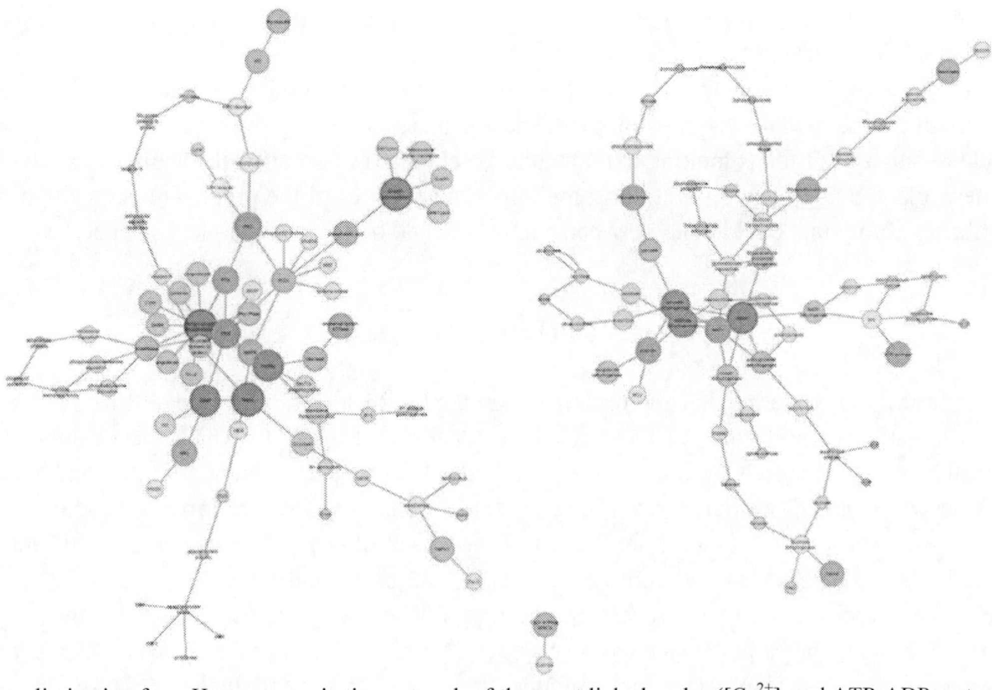

The elimination from Human capacitation network of the most linked nodes ($[Ca^{2+}]_i$ and ATP-ADP system) caused the collapse of network structure. The node size is proportional to the connection number and the node color gradient is dependent from the closeness centrality. This parameter is computed as: $C_c(n) = 1/avg(L(n,m))$, were L (n,m) is the length of the shortest path between two nodes n and m. The closeness centrality of each node ranges from 0 (light gray) to 1 (dark gray) and it is a measure of how fast the information spreads from a given node to the other nodes. The spatial network arrangement is obtained by using the Cytoscape Spring-embedded Layout (see the text for explanation).

Figure 2. Diagram showing the effect of the elimination from Human capacitation network of the most linked nodes.

The functional importance of hubs in determining the physiology of sperm cell is pointed out by the effect of their removal from the capacitation network. In fact, as predicted by the scale free topology, when two random selected nodes have been eliminated from the capacitation network, the variations on its topology were negligible. On the contrary, when the nodes representing the $[Ca^{2+}]_i$ and the ATP-ADP system have been eliminated, the network structure collapsed (Figure 2).

Reasonably this particular architecture could offer an important evolutionary advantage: the robustness against random failure. In fact a random perturbation will involve the most frequent class of nodes, i.e. the less connected ones, with negligible consequences on network structure. In our model we identified 6 hubs, thus it exist only a probability <5% (6/151) that one of them will be involved in random damages.

In addition, it was possible to identify the nodes with only one link that are, in most cases, the input or output terminal of network. It is also of interest to note that almost half of the nodes (about 45%) have two links, one input and one output edge. This is in agreement with the concept that the networks had a signalling transduction-dedicated structure. On the contrary, terminal events (such as protein phosphorylation or membrane fusion) are highly linked, reasonably because the redundancy of biochemical signalling, as a safety strategy to overlap partial failure of the system.

From these data it emerges that the computational modeling of spermatozoa could be an important tool to explore the physiology of male gametes, inferring important proprieties that could remain out of focus looking at molecular level only. In particular the statistical analysis of networks topology could point out some important features of the process of acquisition of fertilizing ability that could open new perspectives in cell biology and clinical sciences.

THE EVOLUTION OF MODEL

A further step to refine this approach has been the use of specific software able to represent the segregation of molecules in the different sub-cellular districts to which they belong. In particular we used the Cytoscape plug-in Cerebral v2 (http://pathogenomics.ca/cerebral/). As previously described we have realized a database containing the information about boar spermatozoa capacitation. In a further file we defined, as attributes, the intracellular localization of each node. By this way we were able to realize the biological network representing this event as a network of networks (each representing a different sperm cell compartment). This approach had an enormous advantage: it was possible to explore the functional links among plasma membrane, cytosol, acrosome, mitochondria, cytoskeleton and nucleus (Figure 3).

In particular it was found that the node representing "actin polymerization" had some important and unique features:

- it is one of the most connected nodes (see Table 4),
- it links in a specific manner all the intracellular compartments,
- its removal from the network did not affect the global network topology (Table 5 and Figure 4) but caused the loss of five important nodes (and among them the "plasma membrane" and "outer acrosome membrane" fusion).

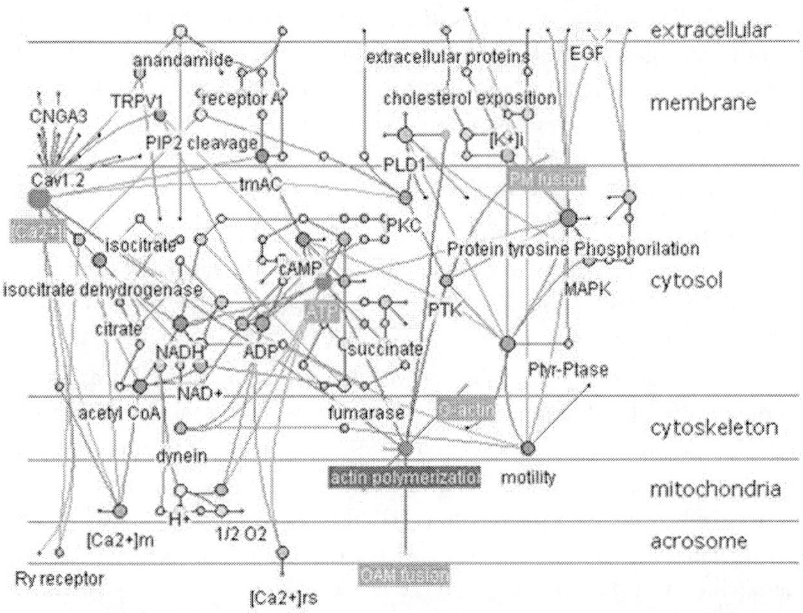

The node size is proportional to the connection number and the node color gradient is dependent from the closeness centrality. This parameter is computed as: $C_c(n) = 1/avg(L(n,m))$, were L (n,m) is the length of the shortest path between two nodes n and m. The closeness centrality of each node ranges from 0 (light gray) to 1 (dark gray) and it is a measure of how fast the information spreads from a given node to the other nodes. All the nodes were localized in their specific subcellular domain (Cytoscape Plugin Cerebral V.2).

Figure 3. Diagram showing the structure of the boar spermatozoa capacitation network and the subcellular localization of nodes.

Table 4. Most connected nodes (the hubs) of boar capacitation network

Node	Number of links
$[Ca^{2+}]_i$	28
ATP	15
Tyr phosphorylation	13
PKA	9
ADP	8
PLD1	8
NADH	8
Actin polymerization	8

It is of great interest, in particular, this last point. In fact, it means that the actin polymerization could behave as a coordinating agent, concurring to the harmonization of different signaling pathways. Importantly this computational model-based hypothesis was experimentally verified. To this aim we incubated boar spermatozoa under capacitating condition either in absence (CTR) or in presence (CD) of 20 μM cytochalasin D, a known specific inhibitor of actin polymerization. Then, we assessed the main signal transduction pathways involved in metabolism of different sub-cellular districts. In particular we monitored:

- *the phospholipase C-γ1 relocalization*: during capacitation PLC- γ1 migrates from cytosol to plasma membrane, where it exerts its enzymatic activity. The biochemical and ultrastructural localization of PLC-γ1 was studied by western blotting or by transmission electron microscopy on CTR and CD treated spermatozoa. As a result it was found that it normally relocalizes from cytosol to the active site (the membrane) independently from CD treatment (see Figure 5A)
- *the changes in protein tyrosine phosphorylation pattern*: as already discussed the pattern of tyrosine phosphorylation changes during the incubation in capacitating condition, in a species specific manner. This parameter was unaffected by the CD treatment. (Figure 5B).
- *the membrane acquisition of chlortetracicline pattern C*: CTC was used to evaluate the completion of membrane remodeling process that occurs during capacitation, in keeping with Mattioli et al. (Mattioli, 1996). Both for CTR and CD samples were assessed at least 200 cells and the percentage of spermatozoa displaying fluorescence pattern C indicative of capacitation (CTC fluorescence over the post acrosomal area) was calculated. As shown in Figure 5C the CD treatment did not affected the kinetic of acquisition of CTC pattern C;
- *the intracellular calcium response to zonae pellucidae*: the calcium probe fluo-3-AM was used to assess the variations in the intracellular calcium concentration in response to the sZP coincubation and it was found that the physiological agonist of AR maintained its ability to stimulate a rapid rise in intracellular calcium concentration also in the CD treated spermatozoa (Figure 5D).

Table 5. Main topological parameters of boar capacitation network before and after actin polymerization removal

Parameter	before	after
N° nodes	153	152
N° edges	204	196
Clustering coefficient	0.056	0.056
Diameter	12	12
Averaged n° neighbours	2.654	2.566
Char. path length	4.995	6.071
	$\gamma = -1.578$	$\gamma = -1.563$
	$r = 0.825$	$r = 0.809$
	$R^2 = 0.849$	$R^2 = 0.898$

The number of nodes represent the total number of molecules involved, the number of edges represents the total number of interaction found, the clustering coefficient is calculated as $CI = 2nI/k(k-1)$,where nI is the number of links connecting the kI neighbours of node I to each other, the network diameter is the largest distance between two nodes, the Averaged n° neighbours represent the mean number of connection of each node, the Char. path length gives the expected distance between two connected nodes. (γ) represents the exponent of node degree law. r and R^2 are the result of correlation and fitting tests.

Taken altogether these data were in perfect agreement with the computational model-based prevision: all the main signaling pathways involved in achievement of capacitation and located in different sub-cellular district were unaffected by CD treatment. As a final confirmation we measured the achievement of fertilizing ability, by inducing the AR with homologous solubilized proteins of zona pellulcida (sZP) (Bernabò et al., 2007; Bernabò et al., 2010c), the physiological inducer or acrosome exocytosis. It was found that the percentage of spermatozoa undergoing spontaneous AR was unaffected by the treatment, while the percentage of spermatozoa able to respond with AR to the ZP coincubation was markedly reduced by the CD administration ($p<0.001$, Figure 6). Then, as inferred, the PM and OAM fusion did not occurred.

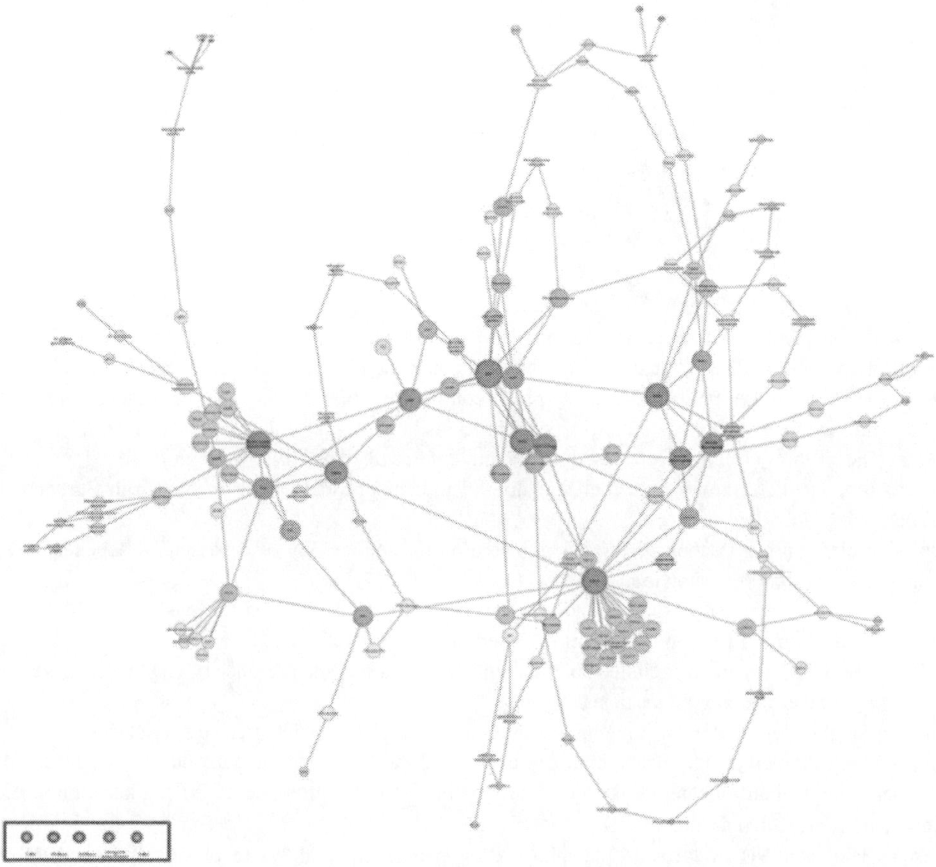

The elimination of "F-actin polymerization" node did not affected the structure of boar sperm capacitation network, but caused the loss of nodes. The node size is proportional to the connection number and the node color gradient is dependent from the closeness centrality. This parameter is computed as: $C_c(n) = 1/avg(L(n,m))$, were L (n,m) is the length of the shortest path between two nodes n and m. The closeness centrality of each node ranges from 0 (light gray) to 1 (dark gray) and it is a measure of how fast the information spreads from a given node to the other nodes. The spatial network arrangement was obtained by using the *Cytoscape Spring-embedded Layout (see the text for explanation).*

Figure 4. Diagram showing the structure of the boar spermatozoa capacitation networkafter "actin polymerization" node removal.

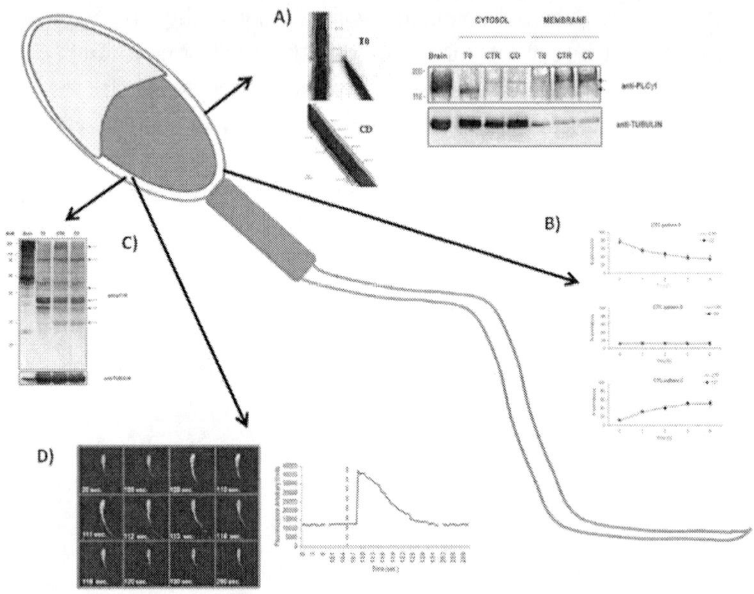

A) Capacitation-dependent PLCγ1 relocalization.

Western Blot analysis of PLCγ1 localization in cytosolic and membrane fractions of freshly ejaculated male gametes (T0) or in spermatozoa incubated under control condition (CTR) or in the presence of CD (CD). The data showing the capacitation-dependent translocation of PLC-γ1 (arrows) from cytosol to membrane. Brain proteins were used as positive control. The filter was normalized on Tubulin expression. The data shown were representative of four independent experiments (from Bernabò et al., 2011).

Transmission electron microscopy pictures demonstrating the topography of capacitation-dependent PLCγ1 relocalization (30.000X) showing:

panel T0): ejaculated spermatozoon displaying few gold particles (arrows) indicative of a low localization of PLCγ1 protein over cell membranes.

panel CD): in vitro capacitated spermatozoon incubated in the presence of CD that localized on its membrane several gold particles (arrows).

B) Kinetic of acquisition of CTC staining pattern A, B and C in CTR and CD treated spermatozoa.

Graphic showing the percentage of spermatozoa displaying the CTC pattern A, B and C, during the 4h of incubation under control conditions (dark continue line) or in the presence of CD (dark dot line). All the values are represented as mean ± SD;

C) Western Blot analysis of tyrosine phosphorylation pattern in total lysate of freshly ejaculated male gametes (T0) or in spermatozoa incubated under control condition (CTR) or in the presence of CD (CD). The arrows indicate the capacitation-related changes in P-Tyr. Brain proteins were used as positive control. The filter was normalized on the Tubulin expression.

D) An example of confocal image gallery of CD treated spermatozoa loaded with Fluo-3AM exposed to sZP (vertical red line). Notice that the rise in the $[Ca^{2+}]_i$ in spermatozoon is evident after <10 sec. from sZP addition, whereas return on the baseline occurs after about 30 sec. (adapted, from Bernabò et al., 2011).

Figure 5. Pathways examined to test the computational-based hypothesis.

This is of great importance for two reasons. As first, the data until now available attribute to the actin dynamics a merely structural role during capacitation: the actin network appearing during this process acts as a diaphragm between the PM and OAM avoiding their fusion. At the moment of AR this physical barrier is rapidly removed and the fusion between the two membranes can happen. If this model was correct it was reasonably to expect that the treatment of spermatozoa with CD, in early phases of capacitation, must leads to the increase in the percentage of spermatozoa undergoing AR because of the inefficiency of PM and OAM separation. In our experiments, on the contrary, it was found that, even in the absence of this diaphragm, the PM and OAM did not fuse. Thus, it possible to hypothesize that in spermatozoa, as in other cells, the role of actin cytoskeleton overcomes this structural function, having a more complex role. Particularly in the light of results of computational experiment it is possible to propose that this event could play an important role in coordination of information flow. This supposition is in keeping with the newly emerging evidence that, in several cellular systems, the cytoskeleton has a key role in signal transduction. As proposed by Janmey, "independent of its mechanical strength, the filaments of the cytoskeleton form a continuous, dynamic connection between nearly all cellular structures, and they present an enormous surface area on which proteins and other cytoplasmic components can dock" (Janmey, 1998). This concept is in accordance with the observation that plasma membrane surface area of a 20-μm-diameter generic cell is on the order of 700 μm^2, in contrast, the total surface area of a typical concentration of 10 mg/ml F-actin is 47,000 μm^2 (Janmey, 1998) and that the diffusion along cytoskeleton tracks could be a reliable alternative to other established ways of intracellular trafficking and signaling, and could therefore provide an additional level of cell function regulation (Shafrir et al., 2000). One implication of this role is that the actin cytoskeleton might provide a signal transduction route and macromolecular scaffold, which, during capacitation, contributes to the spatial organization of signaling pathways (Forgacs et al., 2004). Secondly, it is evident that the *in silico* experiments results were in perfect agreement with those from the *in vitro* approach.

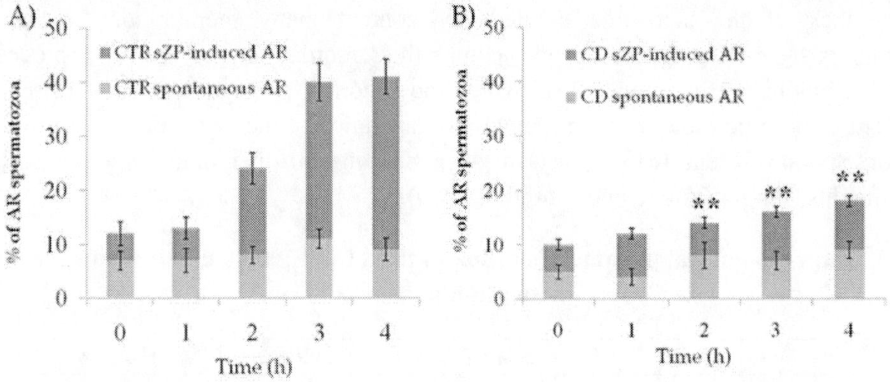

All the values are represented as mean ± SD. ** = p<0.01 vs. CTR, ANOVA test (from Bernabò et al., 2011).

Figure 6. Histogram representing the percentage of spermatozoa undergoing spontaneous and sZP-induced AR in CTR (A) and CD (B) treated spermatozoa.

FROM SEA URCHIN TO WORMS AND HUMAN: SEARCHING THE CONNECTION

Since the computational modeling of male gametes gave positive results, we used it to compare the architecture of biological networks representing the process of sperm cells activation in animals showing a completely different ecology and reproductive behavior. In fact a great advantage of this kind of approach is that the analytical findings are expressed as objective, repeatable and comparable values. As reference we used sea urchin, Caenorhabditis elegans and Human.

Sea urchins are members of the Phylum Echinodermata, they are dioecious, and have five gonads (fivefold symmetry) (Barnes, 1992). The testes are lined with muscles underneath the peritoneum, thus allowing the animal to squeeze its gametes into the surrounding sea water, where fertilization takes place (Barnes, 1992). Spermatozoa of sea urchin became motile only once they contact sea water and are attracted by chemoattractant molecules dispersed by the homologous oocytes the sperm-activating peptides (SAPs) (Neill and Vacquier, 2004). When sperm encounter the egg jelly, the exocytosis of the acrosomal vesicle, the acrosome reaction (AR), occurs and the pH_i-dependent polymerization of actin leads to the extension of an acrosomal tubule, which exposes a new bindin-covered membrane which will fuse with the egg (Barré et al., 2003; Kamei and Glabe, 2003).

C. elegans is an unsegmented, 1 mm long, vermiform and bilaterally symmetrical worm, living in the ground that has hermaphrodites and males. In particular the 99.95% of the individuals are hermaphrodite and the 0.05% are males. Males have a single-lobed gonad, a *vas deferens*, and a tail specialized for mating (L'Heranault, 2003). Hermaphrodites have two ovaries, an oviduct, a chamber where oocytes are fertilized by sperm (the so called "spermatheca") and a single uterus. Its eggs are laid by the hermaphrodite and, after hatching, they pass through four juvenile stages (L1–L4). Hermaphrodites produce sperm at L4 stage (150 sperm per gonadal arm), then they start to produce oocytes. The male can inseminate the hermaphrodite, which will preferentially use male sperm (both types of sperm are stored in the spermatheca). C. elegans spermatozoa are characterized by the absence of the flagellum and the lack of the acrosome, while they content many membranous vesicles, the Membranous Organelles (MOs). They acquire their motility at the end of a process, the spermiogenesis or sperm activation, in which the spherical spermatid extends a pseudopod. This event is made possible by an important reorganization of the cytoskeleton, constituted by a Major Sperm Protein (MSP) instead of actin, and of membrane involving specific microdomains (Fraire-Zamora and Cardullo, 2010).

Table 6. Main topological parameters of sea urchin, C. elegans and Human spermatozoa activation networks

	Sea urchin	C. elegans	Human
N° nodes	127	100	151
N° edges	175	132	202
Clustering coefficient	0.023	0.032	0.028
Diameter	23	23	20
Avg. n° neighbours	2.740	2.620	2.662
Char. path length	8.128	7.816	6.546

The number of nodes represent the total number of molecules involved; the number of edges represents the total number of interactions; the clustering coefficient is calculated as $CI = 2nI/k(k-1)$, where nI is the number of links connecting the kI neighbours of node I to each other; the network diameter is the largest distance between two nodes; the Averaged n° neighbours represents the mean number of connections of each node; the Char. path length gives the expected distance between two connected nodes.

In Human the spermatozoa were ejaculated in vagina, then they travel along the female genital tract until they reach the oviduct, where they reside for a relatively long period (from hours to days) thus acquiring the fertilizing ability.

The comparison of main topological parameters of biological networks representing these three organisms are shown in Table 6 and 7 and the most connected nodes are reported in Table 8.

Table 7. Result of power law fitting of IN and OUT sea urchin, C. elegans and Human spermatozoa activation networks

	Sea urchin		C. elegans		Human	
	IN	OUT	IN	OUT	IN	OUT
r	0.998	0.967	0.992	0.971	0.988	0.997
R^2	0.748	0.924	0.866	0.884	0.890	0.828
γ	-1.589	-2.421	-2.067	-2.127	-1.542	-1.993

(γ) represents the exponent of node degree law
r and R2 are the result of correlation and fitting tests.

As it is immediately evident all the networks share the same scale free topology.

From a biological point of view, it is possible to speculate that this particular architecture could offer important evolutionary advantages, as previously discussed:

- the robustness against random failure;
- the signalling-dedicated structure: each node receives an input signal and transfers an output response;
- the efficiency of signal transduction;
- the fastness of signalling.

It not surprising that the most linked node was, in all the three cases, the intracellular calcium concentration, in fact this ion is a well known ubiquitous second messenger. In sea urchin Ca^{2+} enters the sperm cell through voltage dependent channels (Cav1.2 or 1.3) or through cAMP or cGMP (SpHCN1 or 2) gated channels and it is involved in the control of cAMP, cGMP and in the PIP2/IP3 pathway (Darszon et al., 2005). In C. elegans $[Ca^{2+}]_i$ is controlled by membrane channels and by sequestering the ion in intracellular stores and it concurs in modulating PIP2/IP3 signaling pathways and in regulating the MOs fusion and the onset of motility (Fraire-Zamora and Cardullo, 2010). In Human, as well in other mammals, the capacitation is a calcium-dependent process (as already reported).

The other most connected nodes belong to different classes:

- molecules involved in energetic balance, such as ATP;
- $[H+]_i$: in sea urchin and C. elegans the pH is a key element in control of many biochemical events, such as the membrane polarization, the rearrangement of cytoskeleton proteins (MSP) and the motility (Fraire-Zamora and Cardullo, 2010; Solzin et al., 2004). In mammals spermatozoa, also, the pH modification is an early event of capacitation (Abou-haila and Tulsiani, 2009);

Table 8. Most connected nodes (the hubs) of sea urchin, C. elegans and Human spermatozoa activation networks

Sea urchin		C. elegans		Human	
Node	N° of links	Node	N° of links	Node	N° of links
$[Ca^{2+}]_i$	19	$[Ca^{2+}]_i$	10	$[Ca^{2+}]_i$	25
cGMP	15	$[H^+]_i$	9	ATP	15
$[H^+]_i$	14	Motility	8	Tyr phosph.	13
cAMP	13	ATP	7	PKA	9
ATP	9	Vesicle fusion	7		
		NADH	7		
		NAD^+	6		
		Pseudopod extension	6		

- other second messenger, such cAMP, cGMP. These two cyclic nucleotides concur in signal transduction by modulating the activity of several membrane channels, kinases, phosphatases and many other molecules.

Conclusion

The biology of male reproduction, and male fertility study in particular, is a topic of always growing interest, for its important implication on our all days life. In this chapter we propose the adoption of an innovative approach to study it, taking in account that male germ cells are in every respect complex systems and that it is possiblc to think at thc ability to fertilize as an emerging propriety of the system, which is unpredictable looking only at molecular level. In particular we adopted a biological networks-based modeling strategy that allows us to form some biologically important conclusions:

- the process leading the spermatozoa to gain their full fertilizing ability has about the same architecture in a large variety of organisms, independently from the kind of fertilization (internal versus external), the organism shape (bilateral versus fivefold symmetry), the sexual behavior (male and female versus hermaphrodites and males), the habitat (marine versus terrestrial organisms);
- this architecture is in agreement with Barabási-Albert model of scale free networks;
- this confers to this biological process several important features: the robustness against random failure and the signalling fastness, rapidity and efficiency.

Ultimately, we suggest that the adoption of this approach will potentially open new perspective in important challenges such as drug discovery, diagnosis and therapy of male infertility.

References

Abou-haila, A; Tulsiani, DR. Signal transduction pathways that regulate sperm capacitation and the acrosome reaction. *Arch Biochem Biophys* 2009, 485, 72-81.

Asano, A; Nelson, JL; Zhang, S; Travis, AJ. Characterization of the proteomes associating with three distinct membrane raft sub-types in murine sperm. *Proteomics,* 2010, 10, 3494-3505.

Austin, CR Activation and the correlation between male and female elements in fertilization. *Nature*, 1951 168, 558-559.

Barabási, AL; Oltvai, ZN Network biology: understanding the cell's functional organization. *Nature Reviews Genetics, 2004,* 5, 101-113.

Barbonetti, A; Vassallo, MR; Cinque, B; Antonangelo, C; Sciarretta, F; Santucci, R; D'Angeli, A; Francavilla, S; Francavilla, F. Dynamics of the global tyrosine phosphorylation during capacitation and acquisition of the ability to fuse with oocytes in human spermatozoa. *Biol Reprod*, 2008, 79, 649-656.

Barboni, B; Bernabo`, N; Palestini, P; Botto, L; Pistilli, MG,; Charini, M; Tettamanti, E; Battista N; Maccarrone, M; Mattioli M. Type-1 Cannabinoid Receptors Reduce Membrane Fluidity of Capacitated Boar Sperm by Impairing Their Activation by Bicarbonate. *PLoS ONE* 2011 6, e23038.

Barnes, RD. Invertebrate Zoology. Holt-Saunders International. 1992.

Barré, P; Zschörnig, O; Arnold, K; Huster, D. Structural and dynamical changes of the bindin B18 peptide upon binding to lipid membranes. A solid-state NMR study. *Biochemistry* 2003, 42, 8377-83786.

Bernabò, N; Berardinelli, P; Mauro, A; Russo, V; Lucidi, P; Mattioli, M; Barboni, B. The role of actin in capacitation-related signaling: an in silico and in vitro study. *BMC Systems Biology,* 2011, 30, 47.

Bernabò, N; Mattioli, M; Barboni, B. A Systems Biology Approach to Understanding Male Infertility. In: *Male Infertility*, Edited by Anu Bashamboo and Kenneth David McElreavey, Published by InTech, 2012.

Bernabò, N; Mattioli, M; Barboni, B. The spermatozoa caught in the net: the biological networks to study the male gametes post-ejaculatory life. *BMC Systems Biology*, 2010a, 4, 87.

Bernabò, N; Pistilli, MG; Mattioli, M; Barboni, B. Role of TRPV1 channels in boar spermatozoa acquisition of fertilizing ability. *Molecular and Cellular Endocrinology* 2010b, 323, 224-231.

Bernabò, N; Tettamanti, E; Pistilli, MG; Nardinocchi, D; Berardinelli, P; Mattioli, M; Barboni, B. Effects of 50 Hz extremely low frequency magnetic field on the morphology and function of boar spermatozoa capacitated in vitro. *Theriogenology*, 2007, 67, 801- 815.

Bernabò, N; Tettamanti, E; Russo, V; Martelli. A; Turriani. M; Mattioli. M; Barboni, B. Extremely low frequency electromagnetic field exposure affects fertilization outcome in swine animal model. *Theriogenology*, 2010, 73, 1293-1305.

Blaustein MP; Lederer WJ. Sodium/calcium exchange: its physiological implications. *Physiol Rev*, 1999, 79, 763–854.

Botto, L; Bernabò, N; Palestini, P; Barboni, B. Bicarbonate induces membrane reorganization and CBR1 and TRPV1 endocannabinoid receptor migration in lipid microdomains in capacitating boar spermatozoa. *Journal of Membrane Biology,* 2010, 238, 33-41.

Breitbart, H. Intracellular calcium regulation in sperm capacitation and acrosomal reaction. *Mol Cell Endocrinol*, 2002, 187: 139 - 144.

Breitbart, H; Cohen, G; Rubinstein, S. Role of actin cytoskeleton in mammalian sperm capacitation and the acrosome reaction. *Reproduction*, 2005, 129, 263-268.

Breitbart, H; Cohen, G; Rubinstein, S. Role of actin cytoskeleton in mammalian sperm capacitation and the acrosome reaction. *Reproduction*, 2005, 129, 263–268.

Breitbart, H; Rubinstein, S; Etkovitz, N. Sperm capacitation is regulated by the crosstalk between protein kinase A and C. *Mol Cell Endocrinol*, 2006, 252, 247-249.

Breitbart, H; Etkovitz, N. Role and regulation of EGFR in actin remodeling in sperm capacitation and the acrosome reaction. *Asian Journal of Andrology* 2011, 13, 106-110.

Brewis, IA; Gadella, BM. Sperm surface proteomics: from protein lists to biological function. *Molecular Human Reproduction,* 2010, 16, 68-79.

Bruckbauer, A; Dunne, PD; James, P; Howes, E; Zhou, D; Jones, R; Klenerman, D. Selective diffusion barriers separate membrane compartments. *Biophysical Journal,* 2010, 99, 1-3.

Chang, H; Suarez, SS. Rethinking the relationship between hyperactivation and chemotaxis in mammalian sperm. *Biology of Reproduction,* 2010, 83, 507-513.

Chang, MC Fertilizing capacity of spermatozoa deposited into the fallopian tubes. *Nature,* 1951, 168, 697-698.

Cohen, G; Rubinstein, S; Gur, Y; Breitbart, H. Crosstalk between protein kinase A and C regulates phospholipase D and F-actin formation during sperm capacitation. *Dev Biol,* 2004, 267, 230–241.

Crick FHC Of Molecules and Men. 1966 University of Washington Press, Seattle, WA, USA.

Darszon, A; Nishigaki, T; Wood, C; Trevino, CL; Felix, R; Felix, R; Beltrán, C. Calcium channels and Ca2+ fluctuations in sperm physiology. *Int Rev Cytol* 2005, 243, 79-172.

Florman HM; Jungnickel MK; Sutton KA. Regulating the acrosome reaction. *Int J Dev Biol,* 2008, 52, 503-510.

Forgacs, G; Yook, SH; Janmey, PA; Jeong, H; Burd, CG. Role of the cytoskeleton in signaling networks. *J Cell Sci*, 2004, 117, 2769-2775.

Fraire-Zamora, JJ; Cardullo, RA. The Physiological Acquisition of Amoeboid Motility in Nematode Sperm: Is the Tail the Only Thing the Sperm Lost? *Mol Repr Dev* 2010, 77, 739–750.

Fraser LR; Umar G; Sayed S. Na(+)-requiring mechanisms modulate capacitation and acrosomal exocytosis in mouse spermatozoa. *J Reprod Fertil*, 1993, 97, 539-549.

Fraser, LR; Abeydeera, LR; Niwa, K. Ca(2+)-regulating mechanisms that modulate bull sperm capacitation and acrosomal exocytosis as determined by chlortetracycline analysis. *Molecular Reproduction and Development,* 1995, 40, 233-241.

Fraser, LR; Herod, JE. Expression of capacitation-dependent changes in chlortetracycline fluorescence patterns in mouse spermatozoa requires a suitable glycolysable substrate. *Journal of Reproduction and Fertility,* 1990, 88, 611-621.

Gadella, BM; Harrison, RA. Capacitation induces cyclic adenosine 3',5'-monophosphate-dependent, but apoptosis-unrelated, exposure of aminophospholipids at the apical head plasma membrane of boar sperm cells. *Biology of Reproduction,* 2002, 67, 340-350.

Gadella, BM; Harrison, RA. The capacitating agent bicarbonate induces protein kinase A-dependent changes in phospholipid transbilayer behavior in the sperm plasma membrane. *Development*, 2000, 127, 2407-2420.

Gadella, BM; Tsai, PS; Boerke, A; Brewis, IA. Sperm head membrane reorganization during capacitation. *International Journal of Developmental Biology,* 2008, 5, 473-480.

Gomez-Cambronero, J; Keire, P. Phospholipase D: a novel major player in signal transduction. *Cell Signal*, 1998, 10, 387-397.

Harrison, RA; Gadella, BM. Bicarbonate-induced membrane processing in sperm capacitation. *Theriogenology*, 2005 63, 342-351.

Hickey, KD; Buhr, MM. Lipid bilayer composition affects transmembrane protein orientation and function. *Journal of Lipids,* 2011, 208457.

Janmey PA. The cytoskeleton and cell signaling: component localization and mechanical coupling. *Physiol Rev*, 1998, 78, 763-778.

Kamei, N; Glabe, CG. The species-specific egg receptor for sea urchin sperm adhesion is EBR1, a novel ADAMTS protein. *Genes Dev* 2003, 17, 2502-2507.

Kotwicka, M; Jendraszak, M; Jedrzejczak, P. Phosphatidylserine membrane translocation in human spermatozoa: topography in membrane domains and relation to cell vitality. *Journal of Membranes Biology,* 2011, 240, 165-170.

Krausz, C. Male infertility: pathogenesis and clinical diagnosis. *Best Practice and Research Clinical Endocrinology and Metabolism,* 2011, 25, 271-285.

L'Hernault, SW. Spermatogenesis. *WormBook* 2006, 20, 1-14.

Mattioli, M; Barboni, B; Lucidi, P; Seren, E. Identification of capacitation in boar spermatozoa by chlortetracycline staining. *Theriogenology,* 1996, 45, 373-381.

Navarro, B; Kinichok, Y; Chung, JJ; Clapham, D. Ion channels that control fertility in mammalian spermatozoa. *Int J Dev Biol* 2008, 52, 607-613.

Neill, AT; Vacquier, VD. Ligands and receptors mediating signal transduction in sea urchin spermatozoa. *Reproduction* 2004, 127, 141-149.

O'Rand, MG; Fisher, SJ. Localization of zona pellucida binding sites on rabbit spermatozoa and induction of the acrosome reaction by solubilized zonae. *Developmental Biology,* 1987, 119, 551-559.

Perry, RL; Naeeni, M; Barratt, CL; Warren, MA; Cooke, ID. A time course study of capacitation and the acrosome reaction in human spermatozoa using a revised chlortetracycline pattern classification. *Fertility and Sterility,* 1995, 64, 150-159.

Ren, D; Xia, J. Calcium Signaling Through CatSper Channels in Mammalian Fertilization. *Physiology* 2010, 25, 65-175.

Shafrir, Y; ben-Avraham, D; Forgacs, G. Trafficking and signaling through the cytoskeleton: a specific mechanism. *J Cell Sci*, 2000, 113, 2747-2757.

Solzin, J; Helbig, A; Van, Q; Brown, JE; Hildebrand, E; Weyand, I; Kaupp, UB. Revisiting the role of H+ in chemotactic signaling of sperm. *J Gen Physiol* 2004, 124, 115-124.

Storey, BT. Mammalian sperm metabolism: oxygen and sugar, friend and foe. *Int J Dev Biol*, 2008, 52, 427-437.

Suarez, SS. Control of hyperactivation in sperm. *Human Reproduction Update* 2008, 14, 647-657.

Urner, F; Sakkas, D. Protein phosphorylation in mammalian spermatozoa. *Reproduction,* 2003, 125, 17–26.

Van Regenmortel, MHV. Reductionism and complexity in molecular biology. *EMBO reports,* 2004, 11, 1016-1020.

Venditti, JJ; Bean, BS. Stabilization of membrane-associated alpha-L-fucosidase by the human sperm equatorial segment. *International Journal of Andrology,* 2009, 32, 556-562.

Visconti, PE. Understanding the molecular basis of sperm capacitation through kinase design. *Proc Natl Acad Sci,* 2009, 106, 66766-66768.

Visconti, PE; Bailey, JL; Moore, GD; Pan, D; Olds-Clarke, P; Kopf, GS. Capacitation of mouse spermatozoa. I. Correlation between the capacitation state and protein tyrosine phosphorylation. *Development,* 1995, 121, 1129-1137.

Wennemuth, G; Babcock, BF; Hille, B. Calcium Clearance Mechanisms of Mouse Sperm. *J Gen Physiol*, 2003, 122, 115–128.

WHO Consultation on the zona-free hamster oocyte penetration test and the diagnosis of male fertility. *International Journal of Andrology* 1996, (Suppl. 6).

Yanagimachi, R. (1994). *Mammalian fertilization.* In: Knobil E, Neill JD editors. *The physiology of reproduction (2nd edition).* New York, USA: Raven Press, 1994.

In: Advances in Computational Modeling Research
Editor: Anna Belya Kora
ISBN: 978-1-62618-065-9

Chapter 3

CONSUMER-CENTRIC AGENT-BASED MARKETING MODELS: A HOW-TO APPROACH

Rosanna Garcia and Paul Rummel
Northeastern University, Softricity, LLC, Boston, MA, US

I. INTRODUCTION

The use of micro-simulations, and in particular the subset known as agent-based marketing models (ABMMs), are beginning to find a following in the marketing sciences. They have been used to model, for example, diffusion of innovations (Garcia 2005; Garcia and Jager 2011; Schwarz and Ernst 2009), marketing mix models (Delre, et al. 2007; North, et al. 2010, Think Vine 2010), supply chain management issues (Julka, Srinivasan, and Karimi, 2002, Julka, Karimi, and Srinivasan 2002; Swaminathan, Smith and Sadeh 1998) and geospatial retail store plans (Heppenstall, Evans, and Birkin 2006; Schenk, Löffler, and Rauh 2007). Consumer packaged goods companies, most notably Procter & Gamble, have also used ABMs to improve managerial decision making (North, et al. 2010; Siebel and Kellam, 2003). Rand and Rust (2011) provide a summary, as of date, of other uses for ABMs in marketing. Despite the growing list of companies using agent-based models (ABMs) and marketing academics focusing on modeling consumer markets, ABMMs have still not become a common methodology used by marketing scientists.

This slow acceptance may be because consumer-centric modeling is fraught with complexities that are typically not inherent in other disciplines that utilize micro-simulations. For example, consumers are fickle decision makers, their purchase behaviors are idiosyncratic, as they are influenced by social interactions, advertising, price reductions, etc. As supply-side factors interact to create dynamic market structures, ABMMs try to capture these dynamics and often fail. Critics of agent-based modeling have come up with two common complaints; "one viewpoint is that ABM does not deal with real data and is therefore only for "toy problems", while another viewpoint is that most agent-based models have so many parameters that they can fit any data and are thus nothing more than "computer games" (Rand and Rust 2011, pg. 182). ABMMs have been described as toy models, black boxes, and lacking in verifiability; therefore, are not a valid methodology for rigorous academic learning.

To quell these critiques, Rand and Rust (2011) put forth guidelines for agent-based modeling in marketing in order to build a concrete foundation from which to develop rigor and replicability across models. In this chapter, we continue to build on their work by focusing on the distinctive issues in the verification and validation phases of ABMM development.

We suggest that a major reason that agent-based models have received little support by marketing scientists is the lack of uniform standards in developing, verifying and validating consumer-centric agent-based marketing models. In fact, there is little consistency itself on the definition of verification and validation for simulation techniques. Usage of these terms in micro-modeling simulations is varied even within the marketing literature. Oreskes, Shrader-Frechette and Belitz (1994) contend that verification is demonstrating 'the truth', whereas validation is 'establishing of legitimacy'. As influenced by Boehm (1981), Midgley, Marks and Kunchamwar. (2007) indicate that verification is ensuring the software correctly implements the conceptual model, or is 'solving the equations right' and validation is 'solving the right equations.' Gonzalez and Barr (2000) define verification as 'the process of ensuring that the intelligent system (1) conforms to specifications and (2) that its knowledge base is consistent and complete within itself (p. 412). They further declare that validation is 'the process of ensuring that the output of the intelligent systems is equivalent to those of human experts when given the same inputs' (p. 412). Garcia, Rummel and Hauser (2007) state that verification establishes whether the model, at a macro-level, captures the intent of the real-world model. Rand and Rust (2011) define verification as the *process* by which the implemented model is shown to correspond to the conceptual model and validation is the *process* by which the implemented model is shown to correspond to the real world. Because there seems to be greatest overlap across studies between the Rand and Rust definitions and that of others, we define verification to be 'alignment with the conceptual model' and validation to be 'alignment with the real world'. The conceptual model can be either based on theory or empirical in nature, whereas the real world phenomenon is typically observed in historical facts or historical data.

In the first part of this chapter we draw on the research of Leeflang and Wittink (2000) who laid out a foundation for building marketing models. We use their foundation to outline an approach for developing ABMs. In the second part of this chapter, we discuss details of the issues that arise in verification, validation and calibration for consumer-centric agent-based models, which we call agent-based marketing models (ABMMs). Although the focus of this chapter is on consumer-centric agent-based models, it is still relevant for any programmer looking to develop a validated model that is meant to be used to guide managerial decision making.

II. Model Development Guidelines

In 1970 MIT Professor, John Little, hypothesized why some marketing models were *not* used. He suggested the following possible answers, which are still relevant today for ABMMs: good models are hard to find; good parameterization is even harder; managers do not understand models; and most models are incomplete. The process of designing and building even the simplest of marketing ABMs can be deceivingly difficult. Since ABMs are often used to illustrate emergent or complex behaviors, every minor design detail or small logic error could potentially have large impacts on model performance. To reduce the uncertainty of developing marketing models, Leeflang and Wittink (2000) proposed eleven steps for robust model-building as shown in figure 1.We review these steps as they pertain to ABMMs.

Step 1. Opportunity Identification. Marketing scientists have a wealth of methodologies to use for studying marketing issues: structural equation modeling, structural models, system dynamics modeling, consumer behavior experiments and an endless list of statistical methods. The first question the agent-based marketing modeler needs to ask is, 'why use an ABMM?' "Here the model builder evaluates whether the use of a model can improve managerial decision making" (Leeflang and Wittink 2000, pg. 11). Rand and Rust note a variety of ways that ABMMs may be useful to the marketing scientist. They promote the primary idea that ABMMs can be used to model adaptive consumer behavioral theories, empirically validate these theories and then expand on these theories through predictive modeling. Other methods, such as system dynamics modeling (Rahmandad and Sterman 2008) and structural models, also model consumer behavior in order to be predictive. However, ABMMs have an advantage over these other methodologies by the very nature of their model structure, that of effortlessly allowing policy evaluations at both a micro-level and at a macro-level, even sometimes permitting simultaneous evaluation at both perspectives. For example, unlike the Bass model (Bass 1969), an ABMM can model individual heterogeneous adoption of an innovation (i.e., timing, preferences, network impact, etc.), which is at the micro-level, and also observe the overall market-place diffusion of the innovation, which is at the macro-level.

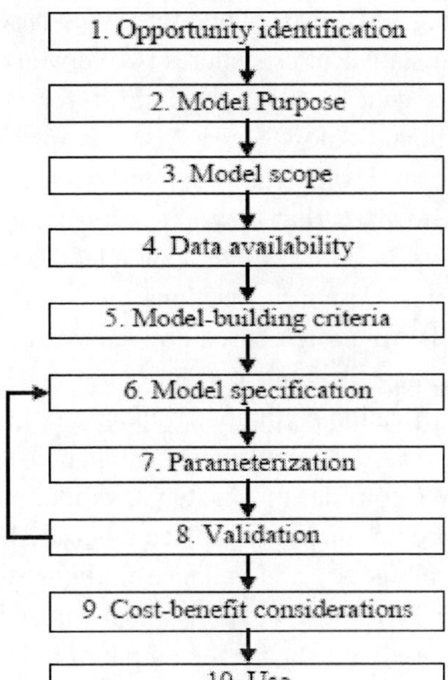

Figure 1. Stages in building an agent-based marketing model (adapted from Leeflang and Wittink 2000).

Because ABMMs can be easily evaluated at multiple levels, we suggest there are two areas in which ABMMs can provide more robust evaluations compared to other methods commonly used in the marketing science. The first is policy testing and the second is forecasting. Policy testings are 'what-if' simulations that help a manager to better understand the market in which they operate. Policy analyses come with their own distinct issues, which ABMMs can overcome. For example, as the Lucas (1976) Critique argues, structural models aren't useful under policy regime changes. "This is because Keynesian models did not take into account the fact that people hold rational expectations, which lead them to change their behavior as soon as the intervention takes place, in order to take advantage of the new policy regime (Fagiolo, Windrum and Moneta 2006, pg. 2013)". The classical macroeconomic model typically used to predict the effects of a new policy would no longer hold because as individuals react to the policy changes, these changes are not accounted for in the model structure. Lucas called for the construction of models that used micro-level behavioral principles to ensure invariant structural relations. ABMMs lend themselves nicely to modeling micro-level behaviors (thereby eliminating the Lucas Critique) and also allowing observation at the macro-level. Policy testing is the focus of ThinkVine (2010), a company who develops marketing mix optimization models using agent-based modeling. In marketing, polices come in the form of new advertising campaigns, couponing, Facebook promotions, etc. Policy implications refer to purchase intent, brand equity, impact on new media, competitor reactions, etc. ABMMs are ideally suited for this purpose of exploring both short-term and long-run implications of various types of policy executions, including profit implications (Leeflang and Wittink 2000).

The second area where ABMMs may provide additional insights compared to other methodologies is in forecasting. ABMMs allow for 'rules-based forecasting,' which is a combination of judgmental and statistical procedures (Collopy and Armstrong 1992). Collopy and Armstrong argue that, in general, there is a need for statistical methods that also incorporate the manager's domain knowledge. ABMMs provide this unique capability because they are 'rules-based', and, when based on empirical input, they allow for statistical analyses. An advantage of ABMMs is that they allow for combining empirical data sets in order to provide better predictive capabilities. For, example, time series data can be combined with vector-autoregressive impulse-response functions for out-of-sample predictions. In early research using ABMMs, models were often based on theory and thus, lacked the ability to be used for prediction and forecasting. It is our current belief that, to date, few ABMMs have demonstrated rigorous verification/calibration/validation necessary to be good predictive/forecasting models. It is our intent with this chapter to suggest how to build this type of rigor into an ABMM. We consider this in step 8, validation, below.

We clarify that ABMMs are not meant to replace other methods of analyzing marketing trends, but are best used to complement and enhance findings resulting from other types of marketing models. See Rand and Rust (2011) and Rahmandad and Sterman (2008) for further details on how ABMMs can be used to enhance the output of other methodologies.

Step 2. Model purpose. In defining a model's purpose, Leeflang and Wittink (2000) state that "in a world flooded by data and increasing tendencies toward customization, there are many opportunities to use models for decision making in marketing" (pg. 11). They identify 'repetitive marketing decisions' as good candidates for model-based automation. This ease of automating a decision making process explains the initial focus of ABMMs on marketing mix modeling, diffusion studies, and supply chain management. However, with the diversity in the model development approaches used by different researchers, there are limited advancements in developing ABM methodologies in marketing, and as such, little success in building models that are useful for repetitive marketing decisions, as suggested by Leeflang and Wittink. For example, despite the popularity of ABMMs to model diffusion, there has been no replicability across models or repetition in model frameworks. Diffusion ABMMs have examined sales prediction (Broekhuizen, Delre and Torres2011), policy implications (Garcia, et al. 2012; Schwarz and Ernst 2009; Zhang, Gensler, and Garcia2011), and network structures (Bohlmann, Calantone and Zhao2010; Delre et al, 2010; Guardiola, et al. 2000; Rahmandad andSterman2008).

As noted above, we suggest that a renewed look at the purpose of an ABMM should focus on policy simulation and forecasting.

Step 3. Model scope. The issue of scope is another dilemma for ABMMs. Numerous studies encourage the linking of marketing decisions to other functional areas, which automatically extends the scope of the model. As these models include more complexity and an integrated systems perspective, they move away from the simplicity encouraged by Little (1970). However, the strength of ABMMs are their ability to model complex systems whereas other marketing models are typically focused on a single agent, i.e., consumer-response, manufacturer- response, competitor-response, etc. Garcia, et al. (2012) is an example of how a model can be complex, yet not capture the complete environment. They explore the diffusion of alternative fuel vehicles by modeling the supply-side with automobile manufacturers' quest for profits, the demand-side with consumer preferences and the role of the governmental policies in impacting market place acceptance of electric vehicles. Yet, they fail to include suppliers and automotive dealers in the model. What should be included within the scope of the model? The answer we believe is that the research question must drive the scope of the model and KISS guides the complexity of the model. KISS refers to keep it simple and straightforward (or similar variations in the vernacular).At the same time, modeling of complex systems requires an awareness for details that may possibly have significant impacts on model behaviors, and should thus not be 'simplified' out of the model."Everything should be made as simple as possible, but not one bit simpler." (attributed to Albert Einstein).

Step 4. Data availability. Leeflang and Wittink (2000) argue the incorporation of detailed marketing data is critical for building marketing models. ABMMs have begun to include empirical data. Garcia, Rummel and Hauser (2007) used conjoint data as the starting point for a diffusion ABMM of screwcaps on fine wines. Broekhuizen, Delre and Torres (2011) empirically validate movie attendance simulation results by conducting a cross-national survey that makes use of the cross-cultural differences in Hofstede's collectivism-individualism index. ThinkVine (2012) uses census data to develop segments of consumers and historical media spends to model firm marketing behaviors. We believe that if ABMMs are to move away from the critique of being a 'toy model', they must be instantiated in empirical data. For the modeler, the problem is not one of availability of data, but more of which type of data to include and at what level. In table 1 below, we list some possible data sources.

An advantage of an ABMM is that it provides a natural way to integrate multiple types of real-world data from various sources. Scaling of the data is the major concern for the programmer, especially if the data is available across different time periods, different levels (micro/macro) and different degrees of detail. Working with 'big data,' such as mobile phone records, can provide a particular challenge to programmers as the quantity of data can quickly bog down a model so that run times become unwieldy. Data management can become a particular challenge for a marketing modeler. Additionally, interpretation of the output becomes complex based on the data used to ground the model. These issues, abundance of data, scaling issues, and interpretation issues, are important reasons why model validation is so critical for ABMM. The validation we propose below demonstrates that the data entered into the model accurately reflects real world phenomenon and provides trust that the model output is useful for managerial decision making.

Table 1. Types of Data Input for ABMMs

• Brand Tracker	• Conjoint analysis	• Competitor analysis
• Shopper Insights	• Regression results	• Twitter data
• Segmentation Research	• Census data	• Governmental Policy results
• Network data	• Historical firm data (i.e., media spending, sales, etc.)	• Primary data specific to study
• Econometric analysis	• Mobile phone records	• Vector autoregressive results
• Sociometric data	• Facebook data	• BASES results

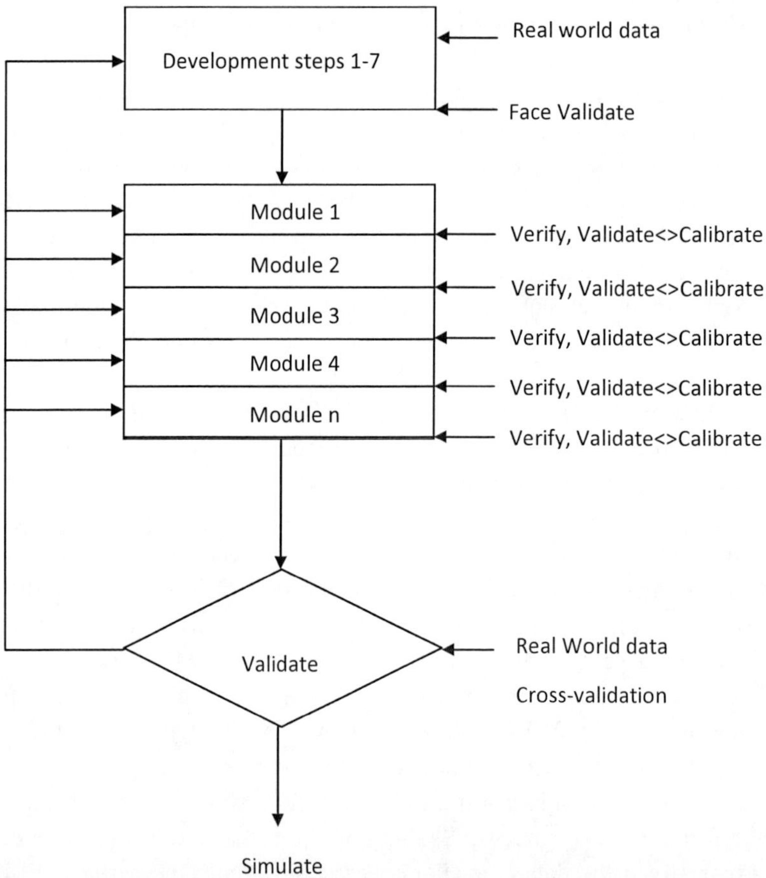

Figure 2. Marketing ABM Design and Build Process.

Step 5. Model building criteria. Model-building criteria relate to model structure, ease of use and implementation strategy (Leeflang and Wittink 2000). ABMMs often lack credibility in these areas. To date we have seen little consistency across model structures, ABMMs remain complex 'black boxes' and thus, fall short in the implementation strategy. In order to address all of these issues, we propose the use of 'modules,' or building blocks for developing ABMMs. This module concept is laid out in figure 2.The modules can best be thought of as a set of tasks that relate to a particular type of agent. For example, in Garcia et al. (2012), there are three types of agents; consumers, manufacturers, and governmental policy setters. The 'consumer module' is programmed around consumer agents making choices as dictated by consumer behavior theory and empirical observations. The 'manufacturer agents' make product design choices based on organization theory and production criteria. The 'government policy-setting agent' makes decision based on social welfare-driven criteria. Taking this modular approach, it becomes easy to see how building each module separately and then combining them together can make a very complicated model less complex. The module structure allows ease of: (1) model development, (2) model validation, (3) model interpretation and (4) verifiability through replicability. The use of modules encourages transparency by removing the black box status as the modules provide model structure, ease of use and implementation strategy.

We list several possible marketing-centric modules to clarify how they can be used to build an ABMM:

- Consumer agents are 'created' by, for example, (1) assignment to a segment, possibly determined by census or survey data; (2) instantiation (grounded) with characteristics using census data; (3) behavioral rules adherence, possibly based on empirical data from surveys results conjoint (Garcia, Rummel and Hauser 2007), regressions/VARs ; (4) network structure assignment possibly based on cell phone, email, or other social media network data; and (5) being presented with rules to follow, as influenced by their environment, possibly based on resource availability, weather, stock markets, political atmosphere, current events, etc…
- Producer agents determine, for example, (1) allocation of consumer stimulus resources (i.e., media spending, marketing mix criteria), attempting to maximize influence on consumers, possibly based on well-established business practices or genetic algorithms to explore promotion spaces; (2) product lines based on engineering or management constraints; (3) relationships with supply chain partners, such as distributors and retailers; (4) production schedules based on internal documents; (5) response to competitor moves; etc.
- Retail agents could be established to determine, for example, (1) inventory of product offerings; (2) relationship with supply chain partners, such as distributors and manufacturers; (3) media spending; etc.

Some of the modules described above set up the initial states of an ABMM's agents (instantiation), while the remainder comprises run-time response rules for the agents. These response rules are typically sensitive to current-time agent micro-level or environment macro-level states.

Instantiation modules are executed at the onset of the model simulation then the run-time modules are executed in an appropriate sequence, yielding a new set of agent states. The appropriate data (by the defined model goals) is collected then run-time modules are repeated.

This run-time module execution and data collection iteration process continues until some appropriate modeled 'end time' is reached. The collected agent-state data is either directly used or further collated and manipulated to extract the behavioral goals of the model. All of these modules are first executed separately and then iterated over appropriate time intervals, yielding the output as defined by the goals of the model. Though any one of the above module layers appears buildable without excessive efforts spent on verifications or validations, it would be highly presumptuous to expect that all layers could be designed and verified en-masse, and even more presumptuous to expect the results of such a model to be believable. Any validating correlation to real-world data or other models would be highly suspect and potentially undefendable.

Step 6. Model specification. Because of the abundance of data available to marketing scientists, it is now possible to specify models at the consumer-level, the household level, the zip-code-level, store-level etc. as well as hourly, daily, weekly, monthly and yearly temporal levels. The question becomes at which level should the model be specified? Model specification is closely related to model scope. Once the scope has been set, model specification should be clearer. We have noted above that scope should be defined by the research question. Additionally, if the model is to follow the KISS guidelines, the model specificity must also be driven by the theoretical model being tested. The theoretical model will define which agents interact with each other, what endogenous factors influence the system and which exogenous factors must be described in order to sufficiently describe the system under study so that the research questions can be addressed adequately. With this focus on theory, agent-rule specificity and KISS, a model should be specified so that the model is defined at a unit of analysis and at a temporal level that allows answering the research question without undue complexity.

Step 7. Parameterization. Marketing scientists have used non-parametric and semi-parametric estimation methods to allow the functional form of main- and interaction-effects to be determined by the data. Some methods include; generalized method of moments (Chintagunta 1992); structural equation models (SEM), instrumental variables in regression analysis (Gasmi, Laffont and Vuong 1992; Kadiyali, Vilcassim and Chintagunta 1998); and Hierarchical Bayes methods (Lenk et al., 1996).Parametric estimations, such as multi-nominal logit and conjoint analysis, are likewise frequently used by marketing scientists. The use of an ABMM does not preclude the use of any estimation method. For example Garcia, Rummel and Hauser (2007) use conjoint analyses within their ABMM. ThinkVine (2012) uses hierarchical Bayes in theirs. As previously noted, the research questions that define the factors of interest in the study should drive model parameterization. An important key for ABMMs is that the parameters be clearly defined, tested through sensitivity analyses, and follow the KISS principle. Transparency in parameter origins will also help reduce the 'black box' critique against ABMMs.

Step 8. Validation. Utilizing ABMs for marketing research most often involves some form of analysis or emulation of people and their consumptive behaviors in an environment full of influences and choices. Whether the marketing ABM is emulative or intellective, multiple verification and validation steps are necessary for model soundness.

As previously noted, the primary goal of this chapter is to provide a standardized process for which to validate an ABMM. Another goal is to demonstrate how an ABMM can be set up as a predictive model for policy analysis and forecasting. Validation has a long history within modeling (Conway, Johnson and Maxwell, 1959), specifically within computational modeling (Carley 1996; Garcia, Rummel and Hauser 2007; Knepell and Arangno, 1993; Rand and Rust 2011). As previously noted, numerous definitions have been put forth for verification, validation and calibration. Based on a literature review, we define verification to be 'alignment with the conceptual model' and validation to be 'alignment with the real world'. We previously did not provide a definition for 'calibration.' Calibration is the process for aligning a model's output to more closely correlate to the targeted validating data or theory. Similar to verification, calibration is a step in the validation process. We delineate the entire validation process in figure 3 and discuss it below.

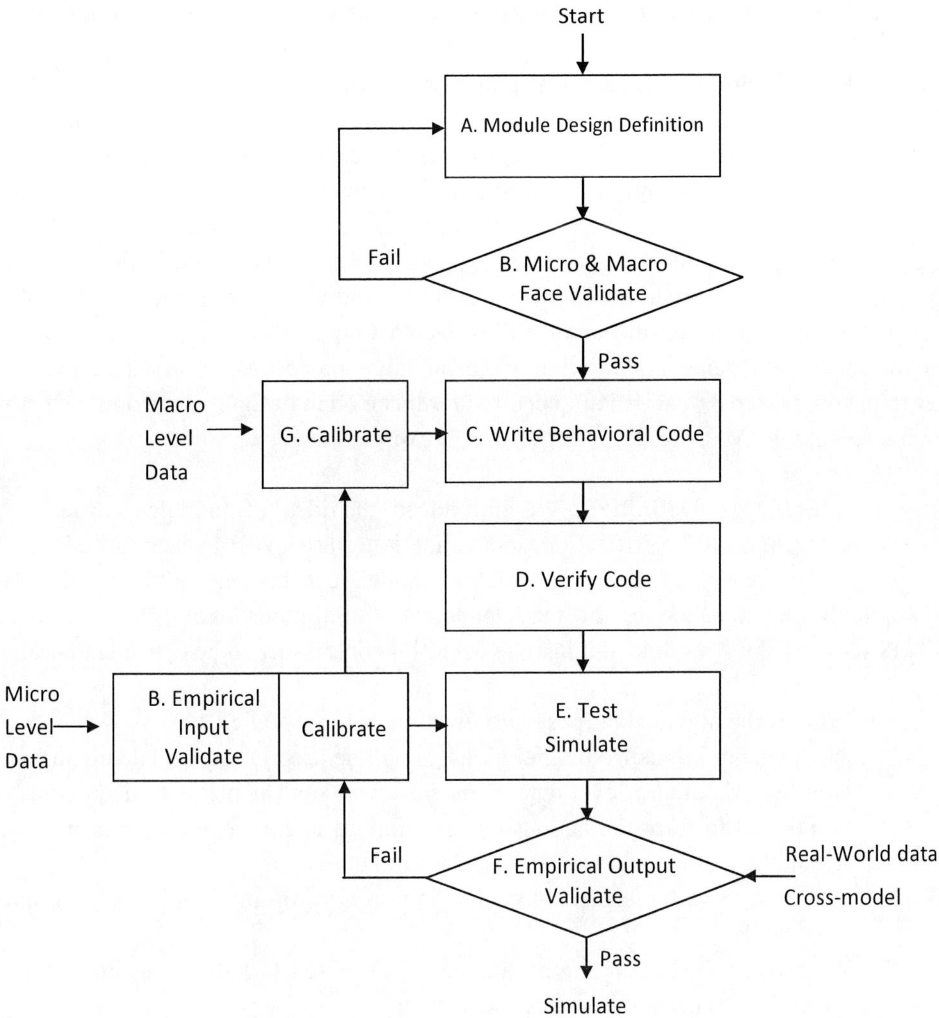

Figure 3. ABMM Module Validation Process.

In explaining this validation process, we use the example of the diffusion of alternative fuel vehicles as detailed in Garcia, et al. (2012). In that ABMM, step 1, 'opportunity identification' presented itself through a National Science Foundation grant on the diffusion of alternative fuel vehicles. Step 2, the 'model purpose,' was to predict the sales of alternative fuel vehicles under different governmental policies. Step 3, the 'model scope,' was constrained to be the direct impact of governmental policies promoting alternative fuel vehicles on consumers and automobile manufacturers. Consumer behavior theory, organizational strategy theory and social welfare theories further defined the scope. Thus, four types of agents were identified: consumer agents, producer agents, automobile design agents, and a single endogenous government agent. Automobile dealers and other partners were not included within the scope. Step 4 'data availability' utilized: (1) a conjoint study of consumer preferences for alternative fuel vehicles, (2) historical data on the sales of automobiles based on fuel type for 2007, the year the conjoint study was conducted, (3) historical data on the impact of Corporate Average Fuel Economy (CAFE) on automobile mpg and retail prices, and (4) AVCEM, Advanced Vehicle Cost and Energy Use Model (Delucchi 2005), an engineering automobile design model. For step 5, the 'model-building criteria', four modules were used to define the model structure: consumer, producer, vehicle design and government policy-setter. Because the research interest was in predicting consumer and producer responses to different governmental policies, step 6, the 'model specification,' was at an individual consumer-choice level for automobiles. The goal of the government agent was to optimize the sales of alternative fuel vehicles, thus, there was no temporal component in that each iteration consumers made a choice decision on a preferred vehicle type, similar to a traditional game theory analysis. Step 7, 'parameterization,' was set by conjoint data for the consumer agents (part-worths for individual preferences) and AVCEM model set the parameters for the automobiles agents (mpg, vehicle weight, vehicle price). Within the context of a general description of the validation process as discussed below, we also demonstrate how in step 8, 'validation', occurred for this model. In figure 3 we outline a process for validating an ABMM using seven steps, A-G. We discuss these steps next.

A. **Module Design Definition.** We introduced the idea of modules earlier in this chapter. Successful verification and validation starts with a clear definition of a module's intended functionality. ABMM models can become quite complex rather quickly and it behooves the modeler to, at a minimum, create clear and concise descriptions of a module, not just the overall model. Issues to be considered are:

    1. What is the intended purpose and function of the module?
    2. What data or inputs from other modules will be used in the module operation?
    3. How will the micro-level data or macro-level data (theories, stylized facts, etc.) be utilized to form the actions or behaviors of agents, environment or stimulus within the module and exogenous to the module?
    4. What are to be the acceptable validation criteria for any empirical input data to the module?
    5. What real-world data or cross-model will be used to validate against and what are to be the acceptable validation criteria?
    6. What data inputs or code parameters will be modified if calibration is needed for validation?
    7. What are to be the defensible levels of data manipulation for possible calibration requirements?

    The clear expression of this information helps the modeler to structure the writing and verification of the code for the module. It also defines how the module is to be finally validated and possibly calibrated. This information becomes an all-important reference for documentation of the model. At a minimum, this information should be embedded within the actual coding of the model as appropriately positioned code 'comments' for debugging. The information should also be placed within a complete-model description document for defense of the model and for anyone who simply needs to learn or later re-learn the workings of the model. These details should be posted on the website, openabm.org, along with the model.

For the alternative fuel vehicle (AFV) policy example, the purpose of the consumer module was to define a consumer choice model as driven by conjoint results; the purpose of the producer module was to define organizational goals, which in this case was profit optimization; the purpose of the vehicle design module was to constrain vehicle design to real world criteria as set in AVCEM (Delucchi 2005); and the purpose of the government policy-setter was to optimize social welfare by decreasing pollution by setting policies impacting automobile choices.

B. Micro Face Validation, Macro Face Validation, and Empirical Input Validation. Similar to Rand and Rust (2011), we incorporate four major points at which validation must occur to consider the model rigorously validated: *micro-face validation, macro-face validation, empirical input validation,* and *empirical output validation* (detailed below in section F). We provide definitions for these below.

Micro-face validation: a test to determine if 'elements of the implemented model (module) correspond "on face" to the real world' (Rand and Rust 2011). How the individual agents are coded to 'think' and 'behave' needs to correspond with real-world examples of what the module is representing. *For the AFV policy example*, micro-face validity occurred at each module. Consumer agents were assigned to a segment using latent class modeling and then checked to see if they were making vehicle choices that seemed 'reasonable' based upon their segment. Producer agents were checked to see if they were making products that sold in the market place given consumer preferences. The vehicle design agents were checked to see if the designs stayed with the engineering constraints set by AVCEM and the government policy-setting agent was checked to determine that it set 'reasonable' policies.

Macro-face validation: a test to determine if 'processes and patterns of the implemented model (module) correspond "on face" to the real world' (Rand and Rust 2011). How the agents respond and behave as a group within the scope of the module needs to correspond with a real-world example of what the module is representing.

For *the AFV policy example,* when combining the modules together, all the agents behaved as expected, with consumer agents choosing their preferred vehicles, manufacturer agents competing against other manufacture agents to design vehicles that maximized firm profits and the government agent setting policies that maximized social welfare.

Empirical input validation: a test to determine if 'data used as inputs to the model corresponds to the real world' (Rand and Rust 2011). Any empirical micro-level data used as input to the module needs to correspond with what the data is representing in the real world. Is the size or distribution of the data appropriate for what it is representing? Does the data represent the proper time period for which the model is expecting? *For the AFV-policy example*, the consumer agents did not pass the test of empirical input validation because 20% of consumer chose alternative fuel vehicles, when in fact the real world AFVs had approximately 3% market share in

2007. This difference in stated preferences (conjoint data) and revealed preferences (sales data) is not a result of programming errors but instead a byproduct of survey response to high uncertainties with new technologies such as AFVs. Conjoint survey respondents are prone to answer more positively than their actual purchase behaviors indicate, especially when the product choice is considered environmentally friendly. However, if the ABMM is to be predictive, the empirical input validation test must be passed. Thus, the consumer module was calibrated as discussed below. The other modules did pass macro-face validation.

C. **Write Behavioral Code.** Much of an ABMM's coding is for the formation of agent behaviors. ABMM *modules* are mostly used and defined for these agent-centered activities and are not usually defined for housekeeping coding tasks like user interfaces, reporting, file management, etc. Regardless of the language used for the model development, ABMM module coding tasks are best started with code 'comments' defined in the module definition, along with any other code structure comments. The more detailed and organized the comments, the easier the actual task of coding. This method of 'comments first' helps keep the coding 'on goal' and reduces the risk of erroneous 'post commenting.' Another set of documentation that assists in model maintenance and re-acquaintance is to keep a record of coding or verification (described below), problems encountered and solutions exacted. Where appropriate, these records can be embedded within the code and/or in the model description document described above in the 'Module Design' section.

   For the AFV-policy example, behaviors were based on both consumer behavioral theory and empirical data. Comments were noted on how theory and data drove the agents' rules.

D. **Verify Code.** As each piece of module behavioral code is written, the first verification step is to simply test the code's execution sufficiently to flush out any coding errors. Secondly, the module should be verified for expected performance, limits and scaling. For example, consider the coding for a consumer that is to have its product attribute propensities instantiated with the micro-level data from a conjoint analysis. Verification of this module could be effected by manipulating either the consumer's attribute propensities or manipulating the targeted product's attributes, and observing if the consumer purchases appropriately. For a more detailed example, consider a competitive analysis of consumer choices of cough-drops. If a particular consumer desires cherry cough-drops over lime, he should end up purchasing cherry over lime. Manipulating that same consumer's propensity for flavor preference should result in the appropriate purchase response. In addition, manipulating the product choice mix of flavors by, for example, eliminating cherry flavors altogether should also result in an appropriately reduced purchase response from that consumer. This type of verification is a part of sensitivity analyses, albeit at a micro-level.

   The limits of the behavior should next be tested by forcing the data input possibilities to extreme values to verify that the resulting consumer agent's behavior is both appropriate and believable. In our cough-drop consumer choice example, this might be done by changing the cherry flavor propensity factor to some extreme high value and observing the purchase response. Does the consumer purchase in a believable and appropriate manner or does the extreme propensity factor cause some odd and possibly unrelated module behavior? What happens to the behaviors when a 'zero' propensity factor is applied? The goal of this part of the verification process is to ensure module robustness even with random data errors.

Lastly, it should be determined if the scaling of the module's output response is proportioned appropriately to other modules that it interacts with. For example, the scaling of how our cherry flavor consumer agent reacts to a product offering might be through the establishment of so many 'purchase probability points' per time interval. The scaling arbitrarily defined as 100 probability points will result in a purchase decision by the consumer, and several other modules may contribute to the final count at purchase decision time. It would be inappropriate to combine this scaling with another module that models the marketing promotion of a producer agent scaled as 'cost savings in pennies' per time interval. The two module outputs can't appropriately be combined in a meaningful manner. All interacting module outputs must be scaled appropriately by operator, magnitude, and time.

For the AFV-policy example, extensive sensitivity analyses were completed. These included, but were not limited to, assigning all consumers to a single segment and seeing how vehicle design tracked to that segment's preferences; setting vehicle design to only one type of vehicle and seeing how consumers responded; setting an astronomically high gas-guzzler tax on gas-combustion vehicles to see how consumers responded. Scaling was also tested by limiting the number of consumers in the market, increasing the number of producers in the market and setting various criteria for recording social welfare costs.

E. **Test Simulate.** Running test simulations upon completion of module verification provides empirical output from which to compare to external validating input. If the defined validation criteria is not met, calibration may be required (described below).In this case, test simulations are executed at each new calibration (modification) point of micro or micro-level data input, followed by another validation test. This iterative calibration process is performed either manually or automatically through coded optimization routines.

F. **Empirical Output Validation.** External validity refers to the adequacy and accuracy of the computational model in matching real world data (Carley 1996). Rand and Rust (2011) define empirical output validation as 'the output of the model corresponding to the real world'. Once the defined goals of the module are realized in code and perform *as designed*, it is prudent to validate the module's performance against external input, whether from a different model or from real-world data. This is accomplished by simply executing the module in a test simulation, recording the output and comparing to the external validation input.

The degree to which an ABMM module should be validated and type of validation used depends on the defined goals of the module. Essentially, the higher the claimed veridicality, the greater the need for more in-depth and higher levels of validation (Carley 1996). As examples, 'pattern validity requires that the pattern of results generated by the computational model matches real patterns of results. Point validity requires that the behavior of the model on each dependent variable, taken one at a time, has the same meaning as the real data. In contrast, distributional validity requires that the distribution of results generated by the computational model has the same distributional characteristics as the real data; e.g., means, standard deviations, and shape of results are the same' (Carley1996). It is important that this external validation not be made with any data used directly by or for instantiation of the module itself, as that would simply be a form of model verification. A common method for utilizing real-world data for both agent instantiation *and* model validation is to first temporally break the data set into two sets, then use the first time set for instantiation, and the later set for validation.

A unique complication of ABMMs is that they are often used to explore potential behaviors in market conditions or environments for which there is no real-world data or existing models to validate against. For example, in 2007 sales of electric vehicles were less than 0.1% of all vehicle sales so not large enough to really measure. In this case, it is essential that EACH module be validated separately. This provides a level of justification for the combination of these modules in a manner that represents the new market environment to be studied.

For the AFV-policy example, empirical output validation was tested by running various governmental policies where historical results were known. For example, tax rebates were in effect in 2007 for hybrid vehicles. Even though the rebates were brand specific since the credit phased out for a particular manufacturer's vehicles once it sold a total of 60,000 eligible vehicles, the overall historical sales of hybrids by vehicle type were attainable for this time period. CAFE had also been in effect, so the model was tested under these various market conditions. Where the model output did not meet the expectations of the historical data, the model was calibrated as discussed below.

G. **Calibration.** When module test simulation outputs do not correlate well enough to be considered 'validated' it is often necessary to 'calibrate' the module. If the micro-level data or macro-level theories used for agent construction/instantiation, mode environment, or stimulus have some latitude for interpretation or margins of error, calibration can be used as a method for bringing a model's output to more closely correlate to the target validating data or theory. Calibration of data used for agent instantiation, environment, or stimulus might involve direct modification of parameters before usage in the module. Calibration of a behavioral theory might involve modifications to behavior factors at the coding level.

Calibration can be performed by modifying the micro-level data or macro-level theory used as input to the module and then measuring it against the module's output or behavior. This single module calibration may only require some simple manual or quickly automated manipulations to input data to obtain the desired validation alignment. For example, if survey data is used as input to instantiate consumer agents with product attribute preferences, a comparison of actual competitive market sales could be used to validate the distribution of the agent's preferences. If this analysis does not show a good enough correlation of the survey data to the historical data, then it might be appropriate to calibrate (modify) the survey data until the module's output indicates a behavior more in line with the 'revealed' preferences of the analysis of real-world data.

Calibration of the complete model with all the modules integrated might possibly require exploration within many data inputs or theory sets used in the model to obtain sufficient validation. This exploration of multi-dimensional data spaces is often performed with the aid of programmed automation. A shotgun approach can always be utilized to explore the results of manipulating the entirety of multi-dimensional data spaces, but when those spaces get to be very large, genetic (or similar) algorithms can be used to speed the process. Alternatively, some commercialized marketing mix optimization software packages validate consumer purchase behaviors against real-world sales data through the manipulation of consumer influencing and instantiation factors, via a 'calibration' process.

Calibration should be used with caution. Calibration can easily be inappropriately utilized as a part of the validation process, becoming a way of 'fudging' a module with inadequate design or data input. If the 'calibrated' theory or data remains within the bounds to which the theory is open to interpretation or the bounds of a dataset's margin of error, then there is a good argument for a level of validation. If these boundaries are excessively large due to the nature of the theory or scarcity of the data, then the measure of validity may be considered too low for the module's defined goals. A solution to this problem would be to supplement the data used in the design of the module rather than 'bending' the existing data. If calibration is deemed necessary, the degree to which the calibrated data's margin of error impacts the module's output margin of error should also be investigated through sensitivity analyses to attain a final measure of validation.

A unique problem does arise in marketing models when historical data is not available for really new or radical technologies. Without historical data, it may not be possible to calibrate an out-of-sample model, which is meant to project future consumer behaviors. In this situation, theory must play a role in calibrating the model.

For the AFV-policy example, the conjoint stated preference results did not align with the historical sales of alternative fuel vehicles. There are several reasons for this misalignment, none that relate to model error. Survey respondents may be more likely to agree to purchase environmentally friendly products and innovative technology. We thus used an iterative process to recalibrate the fuel-type parameters of the conjoint results. First, predicted shares from the consumer module were compared with actual historical shares. If the actual share for a fuel type exceeded the predicted share, the parameters for that alternative (e.g., gas-electric hybrid) were raised. The parameters were lowered if the actual shares were below the predicted (e.g., gas). To adjust the parameters, the following rule (Train 2003) was used:

$$\beta^{1}_{fuel\_type,l} = \beta^{0}_{fuel\_type,l} + \ln\left(\frac{S_{fuel\_type,l}}{\hat{S}^{0}_{fuel\_type,l}}\right)$$

With the new adjusted parameters, shares were predicted and again compared with actual shares. If needed the parameters were adjusted again using the same rule noted above. This calibration was needed for the AFV policy ABMM to be predictive.

Step 9. Cost-benefit consideration. Building a validated ABMM is deceptively difficult. Data management and validation procedures can take many weeks of a modeler's time and then a considerable amount of time for a manager to understand the model output. For the cost-benefit analysis to weigh in favor of developing the model, the model must be able to be used for repetitive decision making. Leeflang and Wittink (2000) argue that standardizing the model so the "essence of repetitive decisions can be captured effectively, or a given model structure has wide applicability, the benefits of the model increase" (pg. 16). With a focus on policy analysis and/or forecasting, ABMMs can garner great benefits to the company willing to invest the initial time and resources into building a standardized model.

Step 10. Use. In this chapter we suggest there are two primary uses for ABMM, for policy analyses and for forecasting prediction. There are a number of different ABMMs that have looked at the impact of policy implementations (Garcia, et al.2012; Broekhuizen, Delre and Torres 2011; ThinkVine 2010). These studies however, have not been replicated. As noted by Leeflang and Wittink (2000) for marketing models to become useful, they must become standardized, replicable, and repetitive. For forecasting, to date, we know of no ABMM that has undergone model development and validation rigorous enough to be considered predictive. However, following the validation process discussed here, ABMMs can demonstrate their ability to be predictive. Future research is needed in this area.

Step 11. Updating. So far as we are aware, there are no meta-analysis studies on ABBMs or studies showing replicability of others' models. Updating does occur with consulting companies, such as ThinkVine, however, their updates are not shared with the public. As previously noted, we believe that for the methodology to mature, replicability must be conducted and reported in such venues as openabm.org. To further a call for transparency and to prevent 'recreating the wheel,' openabm.org, provides a growing collection of tutorials, FAQs on agent-based modeling, a model library intended to provide a locus for authors and modelers to share their models, and forums for modeling-related discussions. By creating a forum for transparency, the 'black box' critique will begin to erode. However, this will not solve the implementation issues of ABMMs. If the role of academics is to advance science, then it up to us as marketing scientists to develop models that are not only easy to use, but also solve meaningful problems of marketing practitioners. The AFV vehicle model by Garcia, et al. (2012) used as the example in this chapter is posted on openabm.org.

Conclusion

In this chapter, we have outlined a standard for developing consumer-centric agent-based models. We argue that the strength of agent-based marketing models (ABMMs) is in policy analysis and forecasting. Using the foundation established by Leeflang and Wittink (2000), we set forth an 11-step process starting with 'opportunity identification' and ending with 'updating.' We elaborated on step 8, 'validation,' to demonstrate a process to undertake in order for ABBMs to be predictive and to perform as managerial decision-making tools. As more ABMMs demonstrate validity in model development, the acceptance of ABMMs in academia and in industry will grow. We also call on marketing researchers to consider building on the work of other scientists in order to demonstrate replicability and repetition across models.

References

Bass, F. M. (1969). "A new product growth model for consumer durables." *Management Science* 15(January): 215-227.

Broekhuizen, Thijs LJ, Sebastiano A. Delre and Anna Torres. "Simulating the cinema market: How cross-cultural differences in social influence explain box office distributions." *Journal of Product Innovation Management* 28, no. 2 (2011): 204-217.

Boehm, Barry. *Software Engineering Economics*. New York, NY: Prentice-Hall, 1981.

Bohlmann, Jonathan D., Roger J. Calantone and Meng Zhao. "The effects of market network heterogeneity on innovation diffusion: An agent-based modeling approach." *Journal of Product Innovation Management* 27, no. 5 (2010): 741-760.

Carley, Kathleen M. *"Validating computational models."* Department of Social and Decision Sciences, Carnegie Mellon University (1996).

Chintagunta, P.K., "Estimating a multinomial probit model of brand choice using the method of simulated moments." *Marketing Science* 11, (1992): 386-407.

Collopy, Fred and J. Scott Armstrong. "Rule-based forecasting: Development and validation of an expert systems approach to combining time series extrapolations." *Management Science* 38.10 (1992): 1394-1414.

Delre, Sebastiano Alessio, Wolfgang Jager, Tammo Bijmolt and Marco A. Janssen, M. A. "Targeting and timing promotional activities: An agent-based model for the takeoff of new products." *Journal of Business Research* 60.8 (2007): 826-835.

Delre, Sebastiano A., Wander Jager, Tammo H.A. Bijmolt and Marco A. Janssen. "Will it spread or not? The effects of social influences and network topology on innovation diffusion." *Journal of Product Innovation Management* 27, no. 2 (2010): 267-282.

Delucchi, Mark. "AVCEM: Advanced Vehicle Cost and Energy Use Model. Overview of AVCEM." Working Paper (2005).

Fagiolo, Giorgio, Paul Windrum and Alessio Moneta. "*Empirical validation of agent-based models: A critical survey.*" Laboratory of Economics and Management, Sant' Anna School of Advanced Studies (2006): 2006-14.

Garcia, Rosanna. "Uses of agentbased modeling ininnovation/new product development research. *"Journal of Product Innovation Management* 22.5 (2005): 380-398.

Garcia, Rosanna, Bruce Clark, William Rand and Sonja Gensler, "Market-level effects of government environmental policies on competitive responses and consumer dynamics: The case of alternative fuel vehicles," *Northeastern University Working Paper Series* (2012).

Garcia, Rosanna and Wander Jager. "From the special issue editors: Agent-based modeling of innovation diffusion." *Journal of Product Innovation Management* 28.2 (2011): 148-151.

Garcia, Rosanna, Paul Rummel, and John Hauser. "Validating agent-based marketing models through conjoint analysis."*Journal of Business Research* 60.8 (2007): 848-857.

Gasmi, Farid, Jean Jacques Laffont, and QuangVuong."Econometric analysis of collusive behavior in a soft-drink market. *"Journal of Economics & Management Strategy* 1, no. 2 (2005): 277-311.

Guardiola, Xavier, Albert Diaz-Guilera, Conrad J. Perez, Alex Arenas and Mateu Llas. "Modeling diffusion of innovations in a social network."*Physical Review E* 66, no. 2 (2002): 026121.

Heppenstall, Alison, Andrew Evans, and Mark Birkin. "Using hybrid agent-based systems to model spatially-influenced retail markets." *Journal of Artificial Societies and Social Simulation* 9.3 (2006).

Julka, Nirupam, Rajagopalan Srinivasan and I. Karimi. "Agent-based supply chain management—1: framework." *Computers & Chemical Engineering* 26.12 (2002): 1755-1769.

Julka, Nirupam, I. Karimi and Rajagopalan Srinivasan. "Agent-based supply chain management—2: a refinery application." *Computers & Chemical Engineering* 26.12 (2002): 1771-1781.

Kadiyali, Vrinda, NaufelVilcassim and PradeepChintagunta. "Product line extensions and competitive market interactions: An empirical analysis." *Journal of Econometrics* 89, no. 1 (1998): 339-363.

Knepell, Peter L. and Deborah C. Arangno. *Simulation validation: a confidence assessment methodology*. Vol. 15. Wiley-IEEE Computer Society Press, 1993.

Leeflang, Peter S.H. and Dick R. Wittink. "Building models for marketing decisions: Past, present and future." *International Journal of Research in Marketing* 17.2 (2000): 105-126.

Lenk, Peter J., Wayne S. DeSarbo, Paul E. Green and Martin R. Young. "Hierarchical Bayes conjoint analysis: Recovery of partworth heterogeneity from reduced experimental designs." *Marketing Science* 15, no. 2 (1996): 173-191.

Little, John D.C. "Models and managers: The concept of a decision calculus." *Management Science* 16, (1970): B466-B485.

Lucas, Robert E. *"Econometric policy evaluation: A critique."* Carnegie-Rochester Conference Series on Public Policy. Vol. 1.No. 1. 1976.

Midgley, David, Robert Marks, and Dinesh Kunchamwar. "Building and assurance of agent-based models: An example and challenge to the field." *Journal of Business Research* 60.8 (2007): 884-893.

North, Michael J., Charles M Macal, James St. Aubin, Prakash Thimmapuram, Mark Bragen, June Hahn, et al. (2010). "Multi-scale agent-based consumer market modeling." *Complexity*, 15(5), 37–47.

Oreskes, Naomi, Kristin Shrader-Frechette and Kenneth Belitz. "Verification, validation, and confirmation of numerical models in the earth sciences." *Science* 263.5147 (1994): 641-646.

Rahmandad, Hazhir and John Sterman. "Heterogeneity and network structure in the dynamics of diffusion: Comparing agent-based and differential equation models." *Management Science* 54, no. 5 (2008): 998-1014.

Rand, William, and Roland T. Rust. "Agent-based modeling in marketing: Guidelines for rigor." *International Journal of Research in Marketing* 28.3 (2011): 181-193.

Schenk, Tilman A., Günter Löffler and Jürgen Rauh (2007). "Agent-based simulation of consumer behavior in grocery shopping on a regional level." *Journal of Business Research* 60(8): 894-903.

Schwarz, Nina and Andreas Ernst."Agent-based modeling of the diffusion of environmental innovations—an empirical approach." *Technological forecasting and social change* 76.4 (2009): 497-511.

Siebel, Fred and Larry Kellam. The virtual world of agent-based modeling: Procter and Gamble's dynamic supply chain. *Perspectives on Business Innovation*, 9 (2003): 22–27.

Swaminathan, Jayashankar M., Stephen F. Smith and Norman M. Sadeh. "Modeling supply chain dynamics: A multiagent approach." *Decision Sciences* 29.3 (1998): 607-632.

ThinkVine. How we use agent-based modeling in our marketing simulation &planning tool. *Technical white paper*. (2010) http://www.thinkvine.com/.

Train, Kenneth E. *Discrete choice methods with simulation*. Cambridge university press, 2003.

Zhang, Ting, Sonja Gensler, and Rosanna Garcia. "A Study of the Diffusion of Alternative Fuel Vehicles: An Agent-Based Modeling Approach." *Journal of Product Innovation Management* 28, no. 2 (2011): 152-168.

In: Advances in Computational Modeling Research
Editor: Anna Belya Kora
ISBN: 978-1-62618-065-9

Chapter 4

LATTICE-BOLTZMANN SIMULATION OF TRANSPORT PHENOMENA IN AGROINDUSTRIAL BIOSYSTEMS

Dayane C.G. Okiyama and José A. Rabi*

University of São Paulo, Faculty of Animal Science and Food Engineering,
Pirassununga, SP, Brazil

ABSTRACT

Numerical methods are helpful either when mathematical difficulties prevent analytical solutions or when obtaining experimental data becomes inhibited due to safety, technological or economic issues. Computational modeling of food and bioprocesses has continuously increased as its importance has been recognized while suitable numerical techniques have been developed and applied. This chapter is particularly devoted to lattice-Boltzmann method (LBM) simulation of transport phenomena involving agroindustrial biosystems.

One may distinguish LBM from both microscale simulation (e.g., molecular dynamics) and macroscale methods (e.g., finite-differences, finite-elements or finite-volumes) inasmuch as the bulk medium is treated like a collection of fictitious particles occupying and traveling within a discrete space, namely, a lattice. From the computational viewpoint, those particles basically follow two procedures, accomplished in discrete time steps. During "collision", particles bump into each other as they arrive at lattice sites and their velocities are rearranged accordingly. During "streaming", collision results are transported to adjacent sites as particles resume their motion along lattice links. By imposing conservation principles to such particle dynamics, macroscopic medium behavior can be numerically simulated.

This chapter particularly focuses on LBM fundamentals as well as on its application to heat and mass (species) transfer involving agroindustrial biosystems. LBM simulation is addressed for biospecific affinity chromatography, supercritical fluid extraction and blast-cooling of catering meals, which are among authors' current research work on computational modeling.

* E-mail address: jrabi@usp.br

1. Computational Modeling of Agroindustrial Biosystems: Brief Overview

Comprehensive understanding of food and bioprocesses may invoke complex governing equations [1]. Moving or free boundaries, anisotropic properties and composition variability may render further difficulties [2], [3]. The related computational modeling has continuously increased as its importance has been recognized while proper numerical techniques have been developed [4],[5]. This chapter is particularly focused on transport phenomena involving agroindustrial biosystems whose mathematical models, though quite established in the literature, are stiff enough to require numerical methods.

Widespread discretization methods for food and bioprocesses comprise finite differences (FDM), finite elements (FEM) and finite volumes (FVM) [3]. Thanks to its relative simplicity, FDM has been typically applied when physical properties are fairly constant in products [6] or when the later can be approximated by uncomplicated shapes such as slices, blocks, cylinders or spheres [7]. FEM has been employed for materials with variable properties, irregular shapes or non-trivial boundary conditions [8]-[11] while FVM has served as basis for CFD (computational fluid dynamics) in the food industry [12]-[14].

Transport phenomena can be numerically simulated on distinct scales [15]. By relying on the continuum concept, the macroscopic approach applies conservation principles to observable properties (e.g., species concentration, temperature, velocity). Their spatial distribution and/or time evolution become governed by differential equations, which have been numerically solved by long-standing discretization methods as those already cited (e.g., FDM, FEM, FVM). On the contrary, the microscopic approach focus on each constituent particle, individually identified together with inter-particle interactions. Newton's law of motion is then applied so as to predict space-time evolution of such colossal particle ensemble. In such a simulation level, referred to as molecular dynamics (MD), large computational effort and resources are required.

Lying between those scales, the mesoscopic approach deals with global effects of particle collections via a distribution function, which attempts to describe their representative behavior. Based on such idea, one numerically implements the lattice-Boltzmann method (LBM), regarded as an extension of its predecessor, namely, lattice-gas cellular automata (LGCA) [16]. The later belongs to a computational modeling technique known as cellular automata, which only allows discrete values for any system property [17].

Computational modeling by means of LBM is more recent if compared to FDM, FEM or FVM [18]. Being a discrete approach for the kinetic theory, LBM has become a promising simulation method for agroindustrial biosystems [19]. Interrelated applications include (but are not restricted to) multiphase displacement [20], packages of agricultural products [21], flow through fibrous materials [22], particle suspensions and particle-fluid interactions [23],[24], agrochemical transport in soils [25],[26], blood fluid dynamics [27],[28], oil reservoirs [29], immiscible fluids [30],[31], liquid-vapor interface [32], chromatography in packed beds [33] and turbulent flows [34].

2. Lattice-Boltzmann Simulation: Basic Concepts

LBM is an alternative bottom-up approach, dissimilar from not only top-down methods (e.g., FDM, FEM or FVM) but also MD simulation, which is a typical bottom-up approach [17]. The contrast from later relies on the fact that LBM treats the macroscale medium, whether solid or fluid, as a collection of fictitious particles occupying a discrete space, namely, the sites of a lattice. According to their speeds, such particles travel between adjacent sites along lattice links. As particles arrive at sites, they mutually collide and are rearranged so as to preserve quantity and momentum. Assuming that such particle dynamics (referred to as streaming and collision) obey conservation principles while being isotropic, bulk medium behavior can be simulated [17], [18].

Inspired on the kinetic gas theory, the underlying idea of LBM is to replace the knowledge on each fictitious particle, in terms of position and momentum, by a description of their overall effect. Such task is mathematically achieved via a distribution function $f = f(\vec{r},\vec{c},t)$ giving the probability of finding, at time t, particles about position $\vec{r}$ with velocities between $\vec{c}$ and $\vec{c}+\mathrm{d}\vec{c}$. Macroscopic (i.e., observable) properties of interest like density ρ, bulk flow velocity $\vec{v}$, internal energy u and pressure p are retrieved from following moments of function f [15],[19]:

$$\begin{aligned} \rho(\vec{r},t) &= \int f(\vec{r},\vec{c},t)\,\mathrm{d}\vec{c} \\ \rho(\vec{r},t)\vec{v}(\vec{r},t) &= \int \vec{c}\, f(\vec{r},\vec{c},t)\,\mathrm{d}\vec{c} \\ \rho(\vec{r},t)u(\vec{r},t) &= \int \tfrac{1}{2}\|\vec{v}_0\|^2 f(\vec{r},\vec{c},t)\,\mathrm{d}\vec{c} \\ \rho(\vec{r},t)[c_s^2\mathbf{I}+\vec{v}(\vec{r},t)\vec{v}(\vec{r},t)] &= \int \vec{c}\vec{c}\, f(\vec{r},\vec{c},t)\,\mathrm{d}\vec{c} \end{aligned} \tag{1}$$

where $\vec{v}_0 = \vec{c}-\vec{v}$ is known as peculiar velocity (= particle velocity $\vec{c}$ relative to flow velocity $\vec{v}$) and pressure is assessed with the help of the sound speed c_s via ideal gas law, $p=\rho c_s^2$. In above equation, **I** is the identity matrix.

The problem (to be computationally solved) then becomes how to obtain function f, which is ruled by Boltzmann's transport equation. In the absence of external forces, such governing differential equation is written as:

$$\frac{\partial f}{\partial t}+\vec{c}\cdot\vec{\nabla}f=\Omega(f) \xrightarrow{\text{BGK approach for } \Omega} \frac{\partial f}{\partial t}+\vec{c}\cdot\vec{\nabla}f=\frac{f^{\mathrm{eq}}-f}{\Delta t_{\mathrm{relax}}} \tag{2}$$

where the so-called collision operator $\Omega = \Omega(f)$ provides the variation rate of function f due to collisions between particles. In the above equation, BGK (Bhatnagar-Gross-Krook) approach has been used in order to linearize the collision operator as $\Omega(f) = (f^{\mathrm{eq}} - f)/\Delta t_{\mathrm{relax}}$, which means that fictitious particle populations tend to local equilibrium values f^{eq} at a rate controlled by a relaxation time $\Delta t_{\mathrm{relax}}$ [35]-[37].

In LBM, Eq. (2) is solved for pre-defined directions (i.e., lattice links), set by the lattice structure. Accordingly, Eq. (2) becomes referred to as lattice-Boltzmann equation (LBE). Distinct structures are indicated as D*n*Q*m*, where *n* is the problem dimensionality (e.g., $n = 1$ = 1-D) and *m* refers to the speed model (= number of distribution functions to be solved).

For simplicity but without loss of generality, let a portion of 1-D lattice structure be considered along *x* direction, as sketched in Figure 1. Such structure comprises a central site linked to two neighbors, where separation distances Δx_k are uniform, being $\Delta x_1 = +\Delta x$ and $\Delta x_2 = -\Delta x$. Fictitious particles may "stream" along links with forward or backward velocities, $\vec{c}_1 = +c\,\hat{x}$ or $\vec{c}_2 = -c\,\hat{x}$, where $c = \Delta x/\Delta t$ is the lattice speed and $\hat{x}$ is a unit vector. If only those two velocities are employed in the model, the lattice is known as D1Q2. Particles at rest can be included by evoking a null velocity ($c_0 = 0$) and the lattice is referred to as D1Q3.

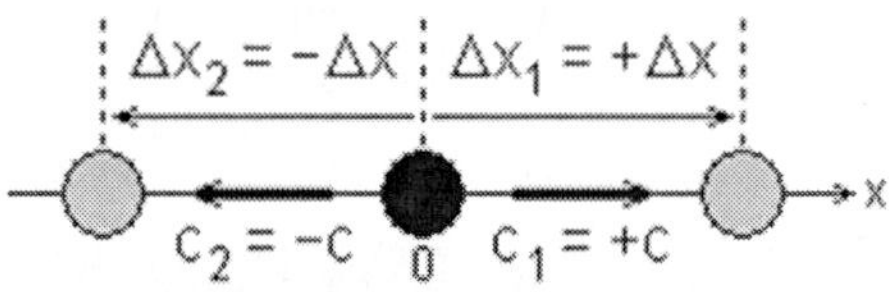

Figure 1. Either D1Q2 or D1Q3 lattice portion for LBM simulations, with lattice distances and speeds.

Since time is discretized in LBM, with Δt as the advancing time step, lattice speeds $c_k = \Delta x_k/\Delta t$ form a discrete set with *k* elements. At this point, it is worth introducing the so-called lattice sound speed c_s, being $c_s^2 = c^2$ for either D1Q2 or D1Q3, whereas $c_s^2 = c^2/2$ for D2Q4 or D2Q5 and $c_s^2 = c^2/3$ for D2Q9, D3Q15 or D3Q19. The aforementioned 2-D and 3-D lattices are depicted elsewhere [15],[18],[19].

Bearing in mind Eq. (2) for a given lattice speed *k*, one may write the following 1-D LBE (under BGK approach) for function $f_k = f_k(x,t)$:

$$\frac{\partial f_k}{\partial t} + c_k \frac{\partial f_k}{\partial x} = \frac{f_k^{\text{eq}} - f_k}{\Delta t_{\text{relax}}} \tag{3}$$

For D1Q3, Eq. (3) is written for $k = 0$, 1 and 2, i.e., for functions f_0, f_1 and f_2. For D1Q2, it is written only for $k = 1$ and 2 since function f_0 is disregarded (or, as introduced ahead, the related weighting factor is null). Space-time discretization of Eq. (3) leads to:

$$\frac{f_k(x,t+\Delta t)-f_k(x,t)}{\Delta t} + c_k \frac{f_k(x+\Delta x_k,t+\Delta t)-f_k(x,t+\Delta t)}{\Delta x_k} = \frac{f_k^{\text{eq}}(x,t)-f_k(x,t)}{\Delta t_{\text{relax}}} \tag{4}$$

By recalling that $c_k = \Delta x_k/\Delta t$ ($\Delta x_k = \pm\Delta x$, according to the streaming direction) while introducing the so-called relaxation parameter $\omega = \Delta t/\Delta t_{\text{relax}}$, one obtains the following algebraic equation:

$$f_k(x + \Delta x_k, t + \Delta t) = [1 - \omega]\, f_k(x,t) + \omega f_k^{\text{eq}}(x,t) \tag{5}$$

Computational evolution of Eq. (5) is performed in two steps. In the collision step (time evolution), each distribution function f_k at each lattice site is updated from instant t to $t + \Delta t$ as:

$$f_k(x,t+\Delta t) = [1-\omega]\, f_k(x,t) + \omega\, f_k^{\text{eq}}(x,t) \tag{6}$$

During the streaming step (space evolution), collision results are propagated to adjacent sites as:

$$f_k(x+\Delta x_k, t+\Delta t) = f_k(x, t+\Delta t) \tag{7}$$

Mesoscale (LBM simulation) and macroscale (observable properties) are connected to one another by the equilibrium distribution function f^{eq} and the relaxation parameter ω. The former is linked to the transport phenomenon (momentum, heat or mass transfer) while the later is linked to the corresponding transport coefficient (kinematic viscosity υ, thermal diffusivity α or species diffusivity D). Indicating such transport coefficient by Γ and the problem dimensionality by n (as before), the following relationship applies [15]:

$$\Gamma = \frac{(\Delta x)^2}{n\,\Delta t}\left(\frac{1}{\omega} - \frac{1}{2}\right) \quad \Leftrightarrow \quad \frac{1}{\omega} = \frac{\Gamma\, n\, \Delta t}{(\Delta x)^2} + \frac{1}{2} \tag{8}$$

In Eq. (8), it goes without saying that $\Gamma = \upsilon$ for momentum transfer, $\Gamma = \alpha$ for heat transfer and $\Gamma = D$ for mass (species) transfer.

As cited above, the equilibrium distribution function f^{eq} is defined in line with the nature of the transport phenomenon to be simulated. For relatively low bulk flow velocity $\vec{v}$ and bearing in mind a given lattice direction k, the following general expression has been proposed [16]:

$$f_k^{\text{eq}} = A_k + B_k(\vec{c}_k \cdot \vec{v}) + C_k(\vec{c}_k \cdot \vec{v})^2 + D_k(\vec{v} \cdot \vec{v}) \tag{9}$$

where A_k, B_k, C_k and D_k are coefficients properly defined [15],[19]. For fluid flow simulations, Eq. (9) assumes the following form:

$$f_k^{\text{eq}} = w_k\, \rho \left[1 + \frac{(\vec{c}_k \cdot \vec{v})}{c_s^2} + \frac{(\vec{c}_k \cdot \vec{v})^2}{2c_s^4} - \frac{(\vec{v} \cdot \vec{v})}{2c_s^2}\right] \tag{10}$$

where w_k are weighting factors fulfilling a normalizing condition $\sum_k w_k = 1$. In order to simulate convective-diffusive transport of any passive scalar $\varphi = \varphi(\vec{r},t)$ (e.g., species concentration or temperature) relying on a known fluid flow $\vec{v}$, the following expression applies:

$$f_k^{\text{eq}} = w_k\,\varphi\left[1+\frac{(\vec{c}_k\cdot\vec{v})}{c_s^2}\right] \tag{11}$$

while for diffusion-dominant transport the equilibrium distribution function reduces to:

$$f_k^{\text{eq}} = w_k\,\varphi \tag{12}$$

For D1Q2, weighting factors are $w_0 = 0$ (i.e., f_0 is disregarded) and $w_1 = w_2 = 1/2$ for each streaming direction (i.e., for f_1 and f_2). For D1Q3, weighting factors are $w_0 = 4/6$ and $w_1 = w_2 = 1/6$. Corresponding values for 2-D and 3-D lattices can be found in [15].

Last but not least, one may retrieve macroscopic (i.e., observable) properties by assessing discrete moments of the distribution functions f_k numerically simulated via LBM, in line with Eq. (1). For instance, in isothermal fluid flow simulations, density ρ and bulk velocity components v_i ($i = x, y, z$) are then assessed as:

$$\rho = \sum_k f_k \quad , \quad \rho v_i = \sum_k f_k\, c_{k,i} \tag{13}$$

while any passive scalar φ (either in convection-diffusion or diffusion-dominant simulations) is retrieved from distribution functions f_k simply as:

$$\varphi = \sum_k f_k \tag{14}$$

3. Agroindustrial Biosystems (and Related LBM Simulation) Addressed in This Chapter

3.1. Introductory Remarks

Agroindustrial biosystems offer a wide horizon for LBM research and by no means has this chapter aimed at an in-depth presentation. Rather, the idea is to focus on LBM simulation of heat and mass (i.e., species) transfer involving agroindustrial biosystems of interest, here referred to as "case studies". Specifically, LBM simulations are presented and discussed for biospecific affinity chromatography (BAC), supercritical fluid extraction (SFE) and blast-cooling of catering meals.

Relying on specific interactions between biomolecules like proteins and enzymes, BAC has become an attractive separation technology, employed with rising frequency [38]. It is a valuable separation technique for biological macromolecules as elevated degree of purification can be achieved after one single process step [39].

Claimed as a promising clean technology for industrial purposes, SFE has been applied to food and natural products, typically with supercritical CO_2 as solvent. Industrial SFE applications comprise for instance (but not restricted to) extraction of essential oils, pigments and aromatic components [40]-[42].

Blast-cooling of recently-cooked catering meals is vital for food safety so that public health authorities have set specific time-temperature combinations to prevent *Clostridium perfringens* outbreaks in meals [43],[44]. As qualitative hazard-based control measures can be more or less efficient than necessary, computational modeling becomes a helpful tool for quantitative risk analysis allied to process improvement and re-engineering [45]-[47].

Dynamic models for those previous food and bioprocesses have been one-dimensional so that LBM simulators here discussed were implemented on D1Q2 lattices. Programming language was FORTRAN standard 90/95 and code lines follow those presented and discussed in [15] for transport phenomena of similar nature. Fluid flow simulation is beyond the scope of this chapter so that fluid velocity is allegedly known *a priori*, wherever required.

3.2. Computational Modeling Issues Concerning BAC and SFE

Despite BAC and SFE are fairly distinct processes, their model frameworks share common aspects. Indeed, typical models for BAC and SFE have relied on 1-D time-dependent variations of species (adsorbate or extract) concentrations, with governing differential equations written in view of axisymmetric cylindrical fixed beds of inner radius R and length L. In other words, such dynamic models have invoked first-order spatial dependence so that species concentrations have been assumed to vary along a single coordinate, besides depending on time.

Species concentrations have then been modeled as $\phi = \phi(x,t)$ and $\theta = \theta(x,t)$, respectively in fluid (= mobile) and solid (= fixed) phases, where x is axial coordinate and t is time. In general, chromatographic columns are vertically oriented while extraction vessels might be horizontal as well. Regardless of such orientation, inlet is at $x = 0$ and outlet is at $x = L$. Porosity ε is allegedly uniform throughout the bed.

In BAC, the flowing fluid is a percolating solution while species is an adsorbate of interest (e.g., protein or enzyme); in SFE, one deals with a supercritical fluid (usually CO_2) and an extract (e.g., essential oil or pigment). If the volumetric flow rate $\dot{V}$ remains constant, the interstitial velocity $v = \dot{V}/(\varepsilon\pi R^2)$ results uniform. In model equations, one may alternatively employ the so-called seepage velocity $\overline{v} = \dot{V}/(\pi R^2)$. Those velocities are related to each other according to Dupuit-Forchheimer relation, namely, $\overline{v} = \varepsilon v \Leftrightarrow v = \overline{v}/\varepsilon$ [48].

Governing equations for fluid-phase species concentration are practically the same for both BAC and SFE inasmuch as convective and diffusive transport have been invoked. In terms of LBM implementation, the presence of any aforesaid transport mechanism in model equations renders (say) straightforward programming of both collision and streaming, Eqs. (6) and (7).

The dissimilarity between BAC and SFE essentially relies on the source (or sink) term concerning the sorption-desorption kinetics, which is directly related to the governing equation for solid-phase species concentration. Though depending on time t, such differential equation has usually lacked terms explicitly depending on axial coordinate x. An interesting issue arises with respect to the spatial evolution inherent to the streaming step so that one may follow two distinct approaches when implementing LBM, [49], namely:

- *Imposition of periodic boundary conditions*: both collision and streaming, Eqs. (6) and (7), are programmed. The system becomes "closed onto itself" since one extremity is only able to "see" the other. From the computational viewpoint, the distribution function f_k leaving one boundary equals that entering the other. For a 1-D lattice with sites identified as $i = 0, 1, ..., N_x - 1, N_x$ (so that $N_x + 1$ is the number of sites), one then imposes $f_1(0) = f_1(N_x)$ and $f_2(N_x) = f_2(0)$ for end points in the streaming step.
- *Suppression of the streaming step*: only collision, Eq. (6), is implemented while streaming, Eq. (7), is simply disregarded. In LBM, it is worth recalling that the former step yields time evolution while the later renders spatial evolution. By dismissing streaming, lattice sites "lose communication" between themselves and, hence, they evolve in time independently from one another. As a result, the number of lattice sites becomes totally irrelevant (e.g., a single site is enough for simulations).

Instead of boundary conditions, systems modeled by zero-order equations with respect to spatial dependence may still "perceive" outside influence via source and/or sink terms. In LBM, such terms are directly inserted into the right-hand side of Eq. (6) [15], so that the collision step becomes extended to:

$$f_k(x,t+\Delta t) = [1-\omega] f_k(x,t) + \omega f_k^{\text{eq}}(x,t) + w_k \dot{S} \Delta t \quad (15)$$

where w_k are weighting factors as already discussed and $\dot{S}$ is a source (> 0) or sink (< 0) term.

Actually, LBM simulation of either BAC or SFE needs two distinct distribution functions "sharing" the same lattice. While $f_k(x,t)$ applies to fluid-phase species concentration, $s_k(x,t)$ refers to the solid-phase counterpart. Bearing in mind the physics underlying BAC and SFE, namely, diffusion-convection in the fluid phase and stationary solid medium, and in line with Eqs (11) and (12), corresponding equilibrium distribution functions f_k^{eq} and s_k^{eq} are defined as:

$$f_k^{\text{eq}}(x,t) = w_k\, \phi(x,t)\left[1 \pm \frac{v}{c}\right] \quad , \quad s_k^{\text{eq}}(x,t) = w_k\, \theta(x,t) \quad (16)$$

where the (+ or –) sign of v/c depends on the streaming direction. Fulfilling the normalizing condition ($\Sigma w_k = 1$), weighting factors w_k are the same for f_k^{eq} and s_k^{eq}. Nevertheless, in view of Eq. (8), relaxation factors ω_{f} and ω_{s} are different for each phase, namely:

$$\text{fluid (axial diffusion): } \frac{1}{\omega_{\text{f}}} = \frac{D}{c\,\Delta x} + \frac{1}{2} \;, \quad \text{solid (no diffusion): } \omega_{\text{s}} = 2 \quad (17)$$

3.3. Computational Modeling Issues Concerning Blast-Cooling of Catering Meals

For food safety purposes, temperature of beef-in-sauce catering meals under blast-cooling has been simulated in response to temperature variations of the cooling air flowing around [50]. In view of the coupling to an existing predictive microbiology model for *Clostridium perfringens* outbreaks in such meals [51], two dynamic thermal models, referred to as TM-0 and TM-1, were proposed and compared aiming at a compromise between realistic description and generality.

As horizontal surfaces of food containers are considerably larger than vertical ones, it is assumed that heat is mostly transferred across bottom and top surfaces of equal areas A. Thermal models differed with respect to the spatial dependence of meal temperatures, which were allowed to depend only on time in TM-0, $T = T(t)$, while they could also vary along a coordinate x normal to heat transfer surfaces in TM-1, $T = T(x,t)$.

Meals are assumed to ultimately transfer heat to cooling air via forced convection and both models invokes a heat transfer coefficient h_c to avoid inherent mathematical difficulties [52]. As described ahead, coefficient h_c belongs to the governing ordinary differential equation in TM-0 whereas it becomes part of the boundary conditions imposed to the governing partial differential equation in TM-1. No distinction is made between coefficients at bottom and top surfaces so that $h_{c,inf} = h_{c,sup} = h_c$ is assumed. For each meal sample investigated in [50] and using either TM-0 or TM-1, such identical value h_c was determined by best-fitting simulated time-temperature profiles for meals against those recorded *in situ*, in response to profiles recorded for the cooling air.

In terms of LBM code, both collision and streaming steps are implemented for TM-1. On the other hand, since spatial variation is not considered in TM-0, one may either impose periodic boundary conditions or suppress the streaming step.

4. Applications: Step-by-Step Case Studies

4.1. Case Study 1: LBM Simulation of Biospecific Affinity Chromatography (BAC)

BAC models tend to be complex enough to justify numerical methods [53],[54]. BAC may benefit from computer modeling if comprehensive knowledge is required [55]. Dynamic 1-D models for BAC have relied on *a priori* known adsorption-desorption kinetics and mass transport by convection and diffusion [54]-[59]. Governing equations for adsorbate concentrations in fluid and solid phases, $\phi = \phi(x,t)$ and $\theta = \theta(x,t)$, have been expressed respectively as:

$$\frac{\partial \phi}{\partial t} + v\frac{\partial \phi}{\partial x} = D\frac{\partial^2 \phi}{\partial x^2} - \frac{1-\varepsilon}{\varepsilon}\dot{r}_{f\to s} \tag{18}$$

$$\frac{\partial \theta}{\partial t} = \dot{r}_{f\to s} = k_{ads}\phi(\theta_{max} - \theta) - k_{des}\theta \tag{19}$$

where v is interstitial velocity, D is adsorbate diffusivity, ε is porosity, k_{ads} and k_{des} are adsorption and desorption coefficients, respectively, and θ_{max} is the maximum adsorption capacity of the chromatographic column. Term $\dot{r}_{f\to s}$ gives the rate at which adsorbate is transferred from fluid to solid phase and thus it behaves as a sink in Eq. (18) and as a source in Eq. (19).

With regard to initial conditions, one may prescribe:

$$\text{at } t = 0\text{: } \phi = 0 \text{ (fluid phase), } \theta = 0 \text{ (solid phase)} \tag{20}$$

As previously remarked, it is worth noting the governing equation for solid-phase concentration θ, Eq. (19), does not explicitly depend on spatial coordinate x so that boundary conditions are only required for Eq. (18). Being $\phi_{in} \neq 0$ a known fluid-phase concentration, one may prescribe the following conditions:

$$\text{at } x = 0 \text{ (inlet): } \phi = \phi_{in}, \text{ at } x = L \text{ (outlet): } \partial\phi/\partial x = 0 \tag{21}$$

Aiming at scale-up procedures, such previous BAC model was cast in dimensionless form. Being $\dot{r}_{ref} \neq 0$ a reference value (defined shortly ahead) for adsorbate transfer rate between fluid and solid phases, dimensionless variables are introduced based on lattice parameters Δx and Δt as well as on process parameters ϕ_{in} and θ_{max} as follows:

$$X = \frac{x}{\Delta x} \quad , \quad \tau = \frac{t}{\Delta t} \quad , \quad \Phi = \frac{\phi}{\phi_{in}} \quad , \quad \Theta = \frac{\theta}{\theta_{max}} \quad , \quad \dot{R} = \frac{\dot{r}_{f\to s}}{\dot{r}_{ref}} \tag{22}$$

As a result, Eqs. (18) and (19) are respectively rewritten in the following dimensionless form:

$$\frac{\partial \Phi}{\partial \tau} + \text{Ma}\frac{\partial \Phi}{\partial X} = \frac{1}{\text{Pe}_\text{m}}\frac{\partial^2 \Phi}{\partial X^2} - \frac{1-\varepsilon}{\varepsilon} P_{max} \dot{R} \tag{23}$$

$$\frac{\partial \Theta}{\partial \tau} = \dot{R} = P_{ads}\,\Phi\,(1-\Theta) - P_{des}\,\Theta \tag{24}$$

Lattice-based Mach number (Ma) and mass-transfer Peclet number (Pe_m) are defined as:

$$\text{Ma} = \frac{v\,\Delta t}{\Delta x} = \frac{v}{c} \quad , \quad \text{Pe}_\text{m} = \frac{(\Delta x)^2}{D\,\Delta t} = \frac{c\,\Delta x}{D} \tag{25}$$

Based on the relationship $\dot{R} = \partial\Theta/\partial\tau$, as suggested in Eq. (24), the reference value for adsorbate transfer rate results as $\dot{r}_{\mathrm{ref}} = \theta_{\max}/\Delta t$ and the remaining dimensionless parameters become:

$$P_{\max} = \frac{\dot{r}_{\mathrm{ref}}\,\Delta t}{\phi_{\mathrm{in}}} = \frac{\theta_{\max}}{\phi_{\mathrm{in}}} \quad , \quad P_{\mathrm{ads}} = k_{\mathrm{ads}}\phi_{\mathrm{in}}\Delta t \quad , \quad P_{\mathrm{des}} = k_{\mathrm{des}}\Delta t \tag{26}$$

By introducing $N_x = L/\Delta x$ so that $N_x + 1$ is the number of lattice sites along x direction (both end points included), boundary conditions imposed by Eq. (21) are cast in dimensionless form as:

$$\text{at } X = 0 \text{ (inlet): } \Phi = 1, \text{ at } X = N_x \text{ (outlet): } \partial\Phi/\partial X = 0 \tag{27}$$

while initial conditions, Eq. (20), assume the following dimensionless form:

$$\text{at } \tau = 0\text{: } \Phi = 0 \text{ (fluid phase)}, \Theta = 0 \text{ (solid phase)} \tag{28}$$

Using separate distribution functions $f_k(X,\tau)$ and $s_k(X,\tau)$ for each phase and a D1Q2 lattice, LBM was implemented according to the dimensionless formulation previously presented. Hence, dimensionless species concentrations in each phase are assessed in view of Eq. (14):

$$\text{fluid: } \Phi(X,\tau) = f_1(X,\tau) + f_2(X,\tau), \text{ solid: } \Theta(X,\tau) = s_1(X,\tau) + s_2(X,\tau) \tag{29}$$

while dimensionless forms of equilibrium distribution functions are based on Eq. (16):

$$\text{fluid: } f_k^{\mathrm{eq}}(X,\tau) = w_k\,\Phi(X,\tau)\,[1 \pm \mathrm{Ma}], \text{ solid: } s_k^{\mathrm{eq}}(X,\tau) = w_k\,\Theta(X,\tau) \tag{30}$$

As commented, sources or sinks are directly introduced on the right-hand-side of Eq. (15). Since adsorbate is transferred from fluid to solid phase, the collision step regarding the fluid phase is extended to include the following sink term in dimensionless form:

$$f_k(X,\tau+\Delta\tau) = [1-\omega_{\mathrm{f}}]\,f_k(X,\tau) + \omega_{\mathrm{f}} f_k^{\mathrm{eq}}(X,\tau) - w_k\left(\tfrac{1-\varepsilon}{\varepsilon}P_{\max}\dot{R}\right)\Delta\tau \tag{31}$$

with $\dot{R} = \partial\Theta/\partial\tau$ assessed with the help of Eq. (24). For the solid phase, a similar expression incorporates a source term, namely:

$$s_k(X,\tau+\Delta\tau) = [1-\omega_{\mathrm{s}}]\,s_k(X,\tau) + \omega_{\mathrm{s}} s_k^{\mathrm{eq}}(X,\tau) + w_k\,\dot{R}\,\Delta\tau \tag{32}$$

In consideration of Eq. (17), distinct relaxation factors are assessed as:

$$\text{fluid phase: } \omega_{\mathrm{f}} = \left(1/\mathrm{Pe_m} + 1/2\right)^{-1}, \text{ solid phase: } \omega_{\mathrm{s}} = 2 \tag{33}$$

As Eq. (24) lacks partial derivatives with respect to axial coordinate X, the related streaming step was simply suppressed, as discussed in the previous section.

Bearing in mind Eq. (28), initial conditions for fluid-phase functions are imposed as:

$$f_1(X,0) = w_1\,\Phi(X,0) \quad , \quad f_2(X,0) = w_2\,\Phi(X,0) \tag{34}$$

and analogously for solid-phase counterparts:

$$s_1(X,0) = w_1\,\Theta(X,0) \quad , \quad s_2(X,0) = w_2\,\Theta(X,0) \tag{35}$$

At bed inlet ($X = 0$), $f_2(0,\tau)$ is obtained via streaming from the adjacent site $f_2(1,\tau)$ so that $f_1(0,\tau)$ is the only unknown. As Eq. (27) imposes $\Phi(0,\tau) = 1$, flux conservation leads to [15]:

$$f_1(0,\tau) = \Phi(0,\tau) - f_2(0,\tau) \;\Rightarrow\; f_1(0,\tau) = 1 - f_2(0,\tau) \tag{36}$$

At bed outlet ($X = N_x$), $f_2(N_x,\tau)$ becomes the unknown as $f_1(N_x,\tau)$ is obtained via streaming from $f_1(N_x-1,\tau)$. By approximating the null Neumann condition, Eq. (27), by means of first-order finite differences, one arrives at the following boundary condition [15]:

$$f_2(N_x,\tau) = f_2(N_x - 1,\tau) \tag{37}$$

LBM simulations were carried out considering a classical work on bioaffinity separation [39], referenced in [56] to test BAC simulators. Along with lattice parameters Δx and Δt, Table 1(*a*) shows data concerning all chromatographic columns investigated in [39], regardless of their length, which were also adopted in simulations performed in [56]. Among possible choices for LBM simulations, the column with length $L = 0.066$ m was considered while bed porosity was set as $\varepsilon = 0.5$ Table 1(*b*) shows the resulting dimensionless parameters and lattice speed results as $c_s = c = \Delta x/\Delta t = 0.002$ m/s.

It is worth citing that species diffusion is disregarded ($D = 0$) in [56] so that Eq. (23) reduces to a 1st-order differential equation. One may then apply numerical methods for initial-value problems (e.g., 4th-order Runge-Kutta method), which tend to be simpler. In the dimensionless model here presented, mass diffusivity D is taken into account via mass-transfer Peclet number $\mathrm{Pe_m}$, Eq. (25). Distinct values were then tested (namely, $\mathrm{Pe_m} = 0.10$, 0.15, 1.00, 10.00 and ∞), the later being implemented by setting $\omega_f = 2$ in line with Eq. (33).

For each testing value for $\mathrm{Pe_m}$ and process parameters as in Table 1, Figure 2(*a*) compares experimental data from [39] with breakthrough curves simulated via LBM. Such curves refer to the fluid-phase adsorbate concentration at bed outlet, i.e., $\Phi_{out}(\tau) = \Phi(N_x,\tau)$. One observes that the slope of simulated breakthrough curve increases as $\mathrm{Pe_m}$ augments, at the expense of larger deviations with respect to the slope of experimental data.

Simulations in [56] were also performed using a 4% lower value for maximum adsorption capacity θ_{max} so that Eq. (26) yields $P'_{max} = 0.96\,P_{max}$. Corresponding LBM simulations are shown in Figure 2(*b*), which suggests that breakthrough curves are shifted to the left, i.e., saturation begins earlier. In addition, tentative 4% and 10% higher values for

θ_{max} were tested and LBM simulations are shown in Figure 2(*c*) and (*d*), respectively. On qualitative basis, breakthrough curve simulated with $Pe_m = 0.15$ combined to 10% higher θ_{max} (= 10% higher P_{max}) seems to better reproduce the experimental data.

Table 1. (*a*) Data concerning chromatographic columns studied in [39], as used for simulations in [56], together with LBM parameters Δx and Δt; (*b*) resulting dimensionless parameters

(*a*) Process parameters	(*b*) Dimensionless parameters
$\phi_{in} = 0.0071\ \text{mol/m}^3$	$P_{max} = \dfrac{\theta_{max}}{\phi_{in}} = 123.24$
$\theta_{max} = 0.875\ \text{mol/m}^3$	$P_{ads} = L\,k_{ads}\,\Delta t = 9.438 \times 10^{-4}$
$k_{ads} = 0.286\ \text{m}^3/(\text{mol}\cdot\text{s})$	$P_{des} = k_{des}\,\Delta t = 2.5 \times 10^{-5}$
$k_{des} = 0.0005\ \text{s}^{-1}$	$\text{Ma} = \dfrac{v\,\Delta t}{\Delta x} = 0.112$
$v = 2.24 \times 10^{-4}\ \text{m/s}$	$\Delta X = 1$
$\Delta x = 0.0001\ \text{m}$	$\Delta\tau = 1$
$\Delta t = 0.05\ \text{s}$	

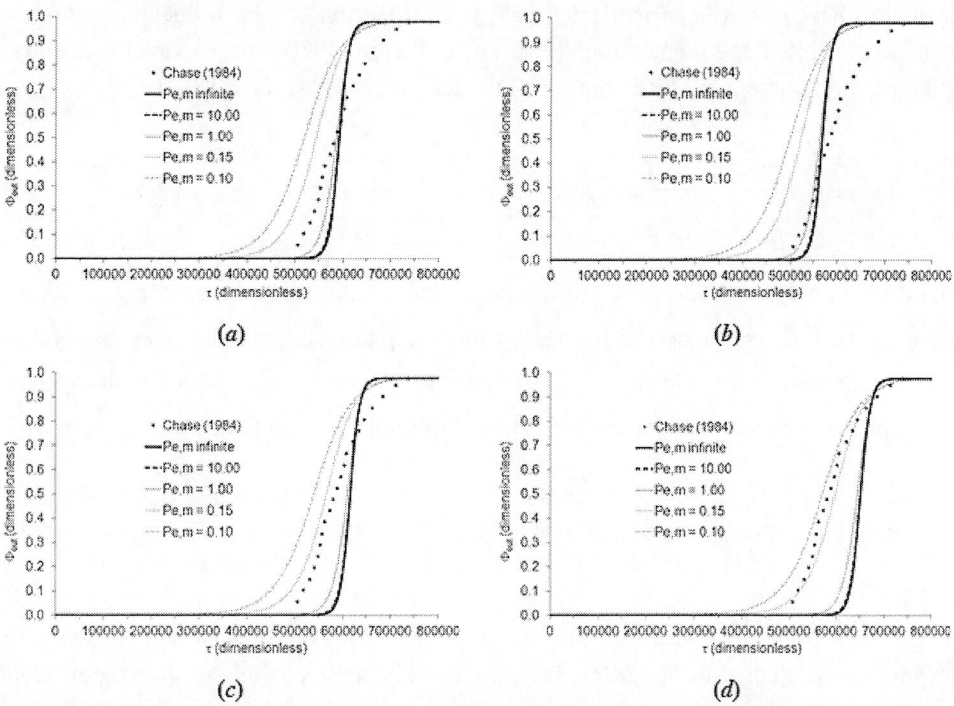

Figure 2. Experimental breakthrough curves from [39] compared to LBM simulations for distinct Pe_m combined to: (*a*) parameters as adopted in [56], (*b*) 4% lower maximum adsorption capacity θ_{max} as suggested in [56], (*c*) 4% higher value for θ_{max}, (*d*) 10% higher value for θ_{max}.

It is worth noting that breakthrough curves simulated via LBM generally reproduce the overall behavior of experimental data. As far as noted discrepancies are concerned, the following points should be pondered. Firstly, porosity was set as $\varepsilon = 0.5$ so as to yield $(1 - \varepsilon)/\varepsilon = 1$ so that Eq. (18) could become equivalent to the governing equation used in [56], which explicitly lack such process parameter. In [56], porosity influence is already taken into accounted in values for θ_{max}, k_{ads} and k_{des}.

Secondly and more importantly, since adsorbate diffusion is neglected in [56] its transport becomes convection-dominant in the fluid phase. Aiming at convective-diffusive transfer, LBM simulations tested distinct mass-transfer Peclet numbers Pe_m. It is worth recalling that numerical schemes for convective transport may yield false diffusion [60],[61] so that breakthrough curves simulated in [56] could be artificially "smoother". LBM is exempt from false diffusion [14] and this may explain why breakthrough curves simulated with values from Table 1 are steeper than experimental data. In addition, $D = 0$ corresponds to $Pe_m \to \infty$ and Figure 2 shows greater deviations precisely for larger Pe_m. Yet, a re-evaluation of some (if not all) parameters in Table 1 is beyond the scope of this chapter.

4.2. Case Study 2: LBM Simulation of Supercritical Fluid Extraction (SFE)

Dynamic 1-D models for SFE have equally relied on *a priori* known adsorption-desorption kinetics and species transport by convection and diffusion [62]. Though the later transport mechanism has been neglected [63]-[67], diffusion can be influential within large equipment, thus justifying its modeling [68]. The governing PDE for the extract concentration $\phi(x,t)$ in the fluid phase is very similar to its counterpart for BAC, Eq. (18), namely:

$$\frac{\partial \phi}{\partial t} + v\frac{\partial \phi}{\partial x} = D\frac{\partial^2 \phi}{\partial x^2} - \frac{1-\varepsilon}{\varepsilon}\dot{r}_{s\to f} \tag{38}$$

While v, D and ε have the same meaning as in BAC model equations, term $\dot{r}_{s\to f}$ gives the rate at which extract is transferred from solid to fluid phase so that it behaves as a source in Eq. (38) rather than a sink. Despite $\dot{r}_{s\to f}$ is identified to $\partial\theta/\partial t$ like in Eq. (19), the governing equation for the extract concentration $\theta(x,t)$ in the solid phase is quite different:

$$\frac{\partial \theta}{\partial t} = \dot{r}_{s\to f} = \frac{1}{t_i}\left(\frac{\phi}{k_P} - \theta\right) \quad , \quad t_i = \mu\frac{l_c^2}{D_i} \tag{39}$$

where k_P is partition (i.e., phase equilibrium) coefficient and t_i is intra-particle diffusion time, which has been evaluated from particle features, namely, a dimensionless shape coefficient μ, a characteristic length l_c and an intra-particle species diffusivity D_i.

If θ_{max} now stands for the maximum extraction capacity of the bed (on concentration basis), one may impose the following initial conditions:

$$\text{at } t = 0\text{: } \phi = 0 \text{ (fluid phase)}, \ \theta = \theta_{max} \text{ (solid phase)} \tag{40}$$

Again, it is worth noting the governing equation for solid-phase concentration θ, Eq. (39), lacks partial derivatives with respect to the spatial coordinate x so that boundary conditions are only required for Eq. (38), which have been imposed as:

at $x = 0$ (inlet): $\phi = 0$, at $x = L$ (outlet): $\partial\phi/\partial x = 0$ (41)

Similar to BAC, SFE model equations were cast in dimensionless form. Let ϕ_{ref} and $\dot{r}_{ref}$ denote reference values respectively for fluid-phase concentration and extract transfer rate from solid to fluid phase, as discussed ahead. Dimensionless variables are introduced as follows:

$$X = \frac{x}{\Delta x} \quad , \quad \tau = \frac{t}{\Delta t} \quad , \quad \Phi = \frac{\phi}{\phi_{ref}} \quad , \quad \Theta = \frac{\theta}{\theta_{max}} \quad , \quad \dot{R} = \frac{\dot{r}_{s\to f}}{\dot{r}_{ref}} \tag{42}$$

Hence, Eqs. (38) and (39) respectively become:

$$\frac{\partial \Phi}{\partial \tau} + \mathrm{Ma}\frac{\partial \Phi}{\partial X} = \frac{1}{\mathrm{Pe_m}}\frac{\partial^2 \Phi}{\partial X^2} - \frac{1-\varepsilon}{\varepsilon} P_{max} \dot{R} \tag{43}$$

$$\frac{\partial \Theta}{\partial \tau} = \dot{R} = P_t\left(\frac{\Phi}{P_P} - \Theta\right) \tag{44}$$

While lattice-based Mach and mass-transfer Peclet numbers, Ma and $\mathrm{Pe_m}$, are defined exactly as in Eq. (25), the remaining dimensionless parameters are:

$$P_{max} = \frac{\dot{r}_{ref}\,\Delta t}{\phi_{ref}} = \frac{\theta_{max}}{\phi_{ref}} \quad , \quad P_t = \frac{\Delta t}{t_i} \quad , \quad P_P = k_P P_{max} \tag{45}$$

Initial and boundary conditions imposed by Eqs. (40) and (41) respectively become:

at $\tau = 0$: $\Phi = 0$ (fluid phase), $\Theta = 1$ (solid phase) (46)

at $X = 0$ (inlet): $\Phi = 0$, at $X = N_x$ (outlet): $\partial\Phi/\partial X = 0$ (47)

At this point, it is worth noting that the null Dirichlet condition at the inlet prevents one from identifying $\phi_{ref} = \phi_{in}$ (as in BAC) since it would lead to division by zero. Without any loss of generality, one may then set $P_{max} = 1$ so that such reference concentration becomes numerically identified to the maximum extraction capacity of the bed, i.e., $\phi_{ref} = \theta_{max}$. As a result, $P_P = k_P$ and dimensionless concentrations Φ and Θ become referenced to the same value.

Using D1Q2 lattice and separate distribution functions $f_k(X,\tau)$ and $s_k(X,\tau)$ for each phase, LBM was implemented based on the previous dimensionless SFE model. Dimensionless species concentrations and equilibrium distribution functions for each phase are assessed through Eqs. (29) and (30), respectively. In view of Eqs. (43) and (44), an additional term must be included in the corresponding collision step like in Eqs. (31) and (32). Yet, as species is extracted from solid to fluid phase, such term becomes a source when implementing Eq. (31) and a sink with respect to Eq. (32). Relaxation factors are assessed as Eq. (33) and the solid-phase streaming step was again suppressed, as already discussed.

Initial conditions for distribution functions in fluid and solid phases are imposed like Eqs. (34) and (35). As far as fluid-phase boundary conditions are concerned, null Neumann condition at outlet is imposed to $f_2(N_x,\tau)$ through first-order finite differences, like in Eq. (37), whereas flux conservation and $\Phi(0,\tau) = 0$ at bed inlet render:

$$f_1(0,\tau) = \Phi(0,\tau) - f_2(0,\tau) \quad \Rightarrow \quad f_1(0,\tau) = -f_2(0,\tau) \tag{48}$$

LBM simulations were performed considering process parameters and experimental results (extraction yield) related to the extraction of essential oil from gorse (*Baccharis trimera*) using supercritical CO_2 [69]. Together with lattice parameters Δx and Δt, Table 2(*a*) shows data from the process carried out at 70°C inside an extractor with length $L = 0.045$ m, radius $R = 0.0055$ m and porosity $\varepsilon = 0.669$. Table 2(*b*) shows the resulting dimensionless parameters. Interstitial fluid velocity results as $v = \dot{V}/(\varepsilon\pi R^2) = 5.243 \times 10^{-4}$ m/s while lattice speed is $c = \Delta x/\Delta t = 0.002$ m/s.

Table 2. (*a*) Data concerning supercritical CO_2 extraction of essential oil from gorse [69] and lattice parameters; (*b*) resulting dimensionless parameters for LBM simulations

(*a*) Process parameters	(*b*) Dimensionless parameters
$\theta_{max} = 27.0\ \text{kg/m}^3$	$P_{max} = \dfrac{\theta_{max}}{\phi_{ref}} = 1$
$k_P = 0.0667$	
$D_i = 2.73\times10^{-10}\ \text{m}^2/\text{s}$	$P_t = \dfrac{\Delta t}{t_i} = 3.276\times10^{-4}$
$l_c = 0.0005$ m	
$\mu_P = 1/3$	$P_P = k_P P_{max} = 0.0667$
$\dot{V} = 1/3\times10^{-7}\ \text{m}^3/\text{s}$	$\text{Ma} = \dfrac{v\,\Delta t}{\Delta x} = 0.262$
$\Delta x = 0.0002$ m	$\Delta X = 1$
$\Delta t = 0.1$ s	$\Delta\tau = 1$

LBM simulations provide dimensionless concentrations in both phases, $\Phi = \Phi(X,\tau)$ and $\Theta = \Theta(X,\tau)$, but experimental results usually refer to the so-called extract yield on mass basis, $Y = Y(t)$. If $\phi_{out}(t) = \phi(L,t)$ is fluid-phase extract concentration at the outlet and m_{total} is total mass of the sample within the bed, extraction yield is assessed as:

$$Y(t) = \frac{1}{m_{total}} \int_0^t \dot{V}\,\phi_{out}(t')\,dt' = \frac{\dot{V}}{m_{total}} \int_0^t \phi(L,t')\,dt' \tag{49}$$

whose dimensionless counterpart is:

$$Y(\tau) = \frac{\dot{V}\,\phi_{\text{ref}}\,\Delta t}{m_{\text{total}}} \int_0^{\tau} \Phi_{\text{out}}(\tau')\,d\tau' = \frac{\dot{V}\,\theta_{\max}\,\Delta t}{m_{\text{total}}} \int_0^{\tau} \Phi(N_x, \tau')\,d\tau' \tag{50}$$

The integration above is numerically implemented with $m_{\text{total}} = 0.015$ kg from [69].

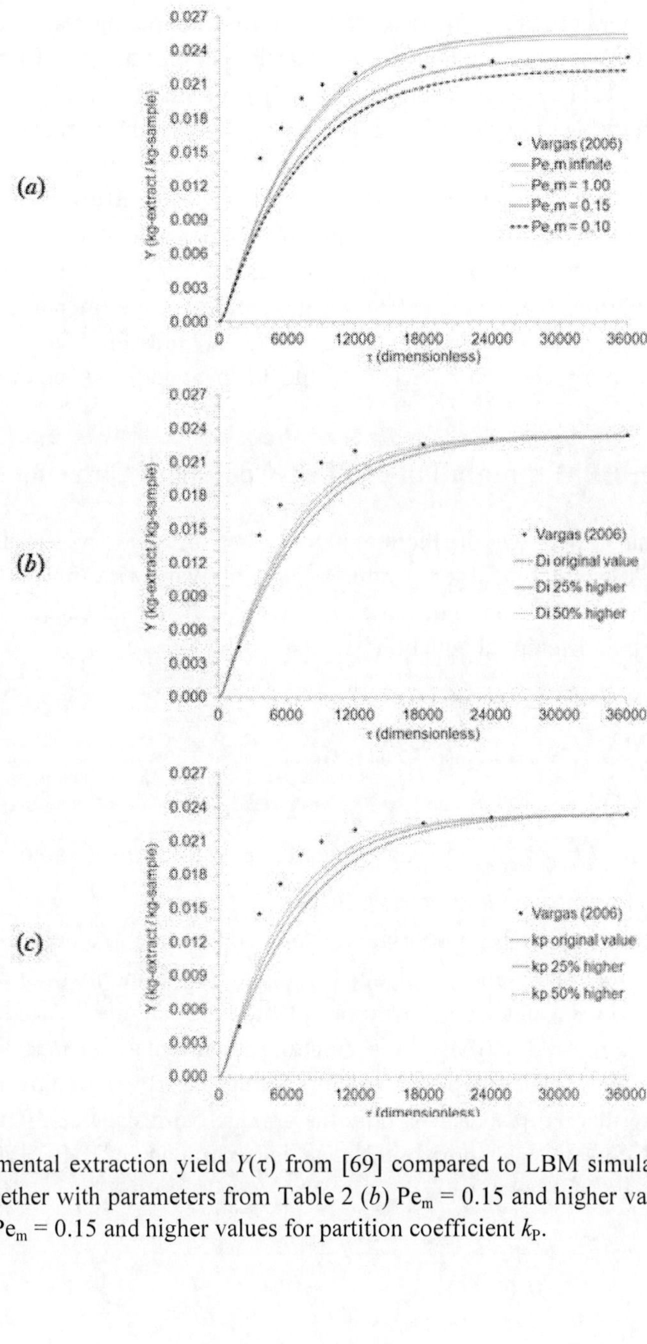

Figure 3. Experimental extraction yield $Y(\tau)$ from [69] compared to LBM simulations for: (*a*) testing values of Pe_{m} together with parameters from Table 2 (*b*) $\text{Pe}_{\text{m}} = 0.15$ and higher values for intra-particle diffusion D_{i}, (*c*) $\text{Pe}_{\text{m}} = 0.15$ and higher values for partition coefficient k_{P}.

In the SFE model proposed by [69], extract diffusion is neglected in the fluid phase. As already commented for BAC, this is equivalent to assume $\mathrm{Pe_m} \to \infty$ in Eq. (43). Conversely, in the dimensionless model here proposed, distinct values for mass-transfer Peclet number were tested, namely, $\mathrm{Pe_m} = 0.10$, 0.15, 1.00 and ∞. The later value is implemented in LBM by setting $\omega_f = 2$, in accordance with Eq. (33). Figure 3(*a*) compares the experimental extraction yield $Y = Y(\tau)$ as measured in [69] against LBM simulations for each testing $\mathrm{Pe_m}$. One observes that $\mathrm{Pe_m}$ dictates the asymptotic value of the extraction yield, with better results for $\mathrm{Pe_m} = 0.15$.

Furthermore, numerical curves diverges somehow from experimental data, although the overall behavior was reasonably simulated. One possible explanation relies on the fact that parameters D_i and k_P (Table 2) were estimated in [69] via a numerical procedure based on a 1st-order model with respect to spatial dependence, i.e., convection-dominant mass transport in the fluid phase. Hence, due to false diffusion effects, such best-fitting values for D_i and k_P might be unsuitable for LBM simulations, which account for mass diffusion as well.

For investigation purposes, tentative 25% and 50% higher values for both D_i and k_P were then tested together with $\mathrm{Pe_m} = 0.15$ and results are respectively presented in Figure 3(*b*) and (*c*). Either way, one verifies that simulated curves get closer to experimental data, which may suggest a proper combination of higher values for D_i and k_P, thus for P_t and P_P via Eq. (45). Again, a re-evaluation of process parameters in Table 2 is beyond the scope of this chapter.

4.3. Case Study 3: LBM Simulation of Blast-Cooling of Catering Meals

Being a preliminary approach, the thermal model of zero-order (TM-0) relies on lumped-parameter analysis [70] so that meal temperatures may only vary with time t, $T_{\mathrm{meal}} = T_{\mathrm{TM\text{-}0}}(t)$. An energy balance for a meal sample of mass m_{meal} and specific heat capacity $c_{\mathrm{p,meal}}$ leads to the following governing differential equation:

$$m_{\mathrm{meal}} c_{\mathrm{p,meal}} \frac{\mathrm{d}T_{\mathrm{TM\text{-}0}}}{\mathrm{d}t} = -\dot{Q}_{\mathrm{meal}\to\mathrm{air,inf}} - \dot{Q}_{\mathrm{meal}\to\mathrm{air,sup}} \tag{51}$$

where $\dot{Q}_{\mathrm{meal}\to\mathrm{air,inf}}$ and $\dot{Q}_{\mathrm{meal}\to\mathrm{air,sup}}$ are heat transfer rates from meal sample to cooling air flowing above and below the container, respectively.

Lumped-parameter analysis is appropriate for low Biot number, namely $\mathrm{Bi} = (h_c\, L_{\mathrm{ref}})/\lambda < 0.1$ [70], being L_{ref} a reference (characteristic) length and λ the thermal conductivity of the meal sample in the problem under consideration. In line with an electrical network analogy [71], such constraint is relaxed in TM-0 by accounting for thermal resistances between meal core and each horizontal surface exposed to the cooling air. By reasoning on the relative magnitude of existing thermal resistances, only the conductive resistance of each half-height $L_{1/2}$ separating meal core to each horizontal edge is accounted in TM-0 [50], so that the previous energy balance leads to:

$$\frac{\mathrm{d}T_{\mathrm{TM\text{-}0}}}{\mathrm{d}t} = -\beta\left[T_{\mathrm{TM\text{-}0}}(t) - T_{\mathrm{air,av}}(t)\right] \tag{52}$$

where $T_{\mathrm{air,av}}(t)$ is an average time-temperature profile for the cooling air, which is assessed from recordings $T_{\mathrm{air,inf}}(t)$ and $T_{\mathrm{air,sup}}(t)$ for the air flowing below and above the container:

$$T_{\mathrm{air,av}}(t) = \frac{T_{\mathrm{air,inf}}(t) + T_{\mathrm{air,sup}}(t)}{2} \tag{53}$$

and the lumped-parameter β (allegedly constant) is given by:

$$\beta = \frac{2A/(m_{\mathrm{meal}} c_{\mathrm{p,meal}})}{1/h_{\mathrm{c}} + L_{1/2}/\lambda_{\mathrm{meal}}} \tag{54}$$

Solution of Eq. (52) requires an initial condition, namely:

$$T_{\mathrm{TM\text{-}0}}(0) = T_0 = T_{\mathrm{exp}}(0) \tag{55}$$

which is assessed from temperatures recorded *in situ* at the start-up of blast-cooling operation. In a conjectural scenario with $T_{\mathrm{air,av}} = T_{\infty}$ = constant (besides assuming that the lumped-parameter β also remains constant), Eq. (52) admits the following analytical solution:

$$T(t) = T_{\infty} + (T_0 - T_{\infty})\exp(-\beta t) \tag{56}$$

Subsequently, a first-order model with respect to spatial dependence (TM-1) was proposed, allowing for one-dimensional variation of meal temperature, besides its time-dependence. In line with the assumption that heat is mostly transferred through horizontal surfaces of meal samples, spatial variation refers to a vertical axis x so that $T_{\mathrm{meal}} = T_{\mathrm{TM\text{-}1}}(x,t)$. An energy balance for a meal sample now leads to the following governing partial differential equation:

$$\frac{\partial T_{\mathrm{TM\text{-}1}}}{\partial t} = \alpha_{\mathrm{meal}} \frac{\partial^2 T_{\mathrm{TM\text{-}1}}}{\partial x^2} \quad , \quad \alpha_{\mathrm{meal}} = \frac{\lambda_{\mathrm{meal}}}{\rho_{\mathrm{meal}}\, c_{\mathrm{p,meal}}} \tag{57}$$

where α_{meal} is the average thermal diffusivity of the catering meal.

As far as the initial condition is concerned, a temperature profile must be prescribed over the meal sample height within the container at $t = 0$:

$$T_{\mathrm{TM\text{-}1}}(x,0) = T_0(x) \tag{58}$$

The simplest option is to impose a uniform profile, i.e. $T_0(x) = T_0$ = constant, being T_0 the same value used in Eq. (55). Equally reasoning on the relative magnitude of thermal resistances as in TM-0, mixed-type (also known as third-type) boundary conditions are imposed at meal bottom ($x = 0$) and meal top ($x = L_{\mathrm{meal}}$) [50]:

$$-\lambda_{\text{meal}} \left.\frac{\partial T_{\text{TM-1}}}{\partial x}\right|_{x=0} = h_{\text{c}}\left[T_{\text{air,inf}}(t) - T_{\text{TM-1}}(0,t)\right]$$
$$-\lambda_{\text{meal}} \left.\frac{\partial T_{\text{TM-1}}}{\partial x}\right|_{x=L_{\text{meal}}} = h_{\text{c}}\left[T_{\text{TM-1}}(L_{\text{meal}},t) - T_{\text{air,sup}}(t)\right] \quad (59)$$

In spite of comprising meat pieces in sauce, meals were treated as single entities whose properties were estimated by weighted-average of separate meat and sauce values, experimentally measured or taken from the literature, as detailed in [50]. Thermophysical parameters directly or indirectly required in TM-0 and TM-1 are shown in Table 3. As an initial approach, such average values for thermophysical properties were assumed to represent the meal sample as a whole and to remain constant during the blast-cooling operation. In addition, simulations were performed in view of a meal sample studied in [50], for which convective heat transfer coefficients were best-fitted as $h_{\text{c,TM-0}} = 8.0$ W/(m^2·K) and $h_{\text{c,TM-1}} = 7.2$ W/(m^2·K), respectively using TM-0 and TM-1.

LBM is applied by assigning a particle distribution function f_k to meal core temperature T, on a D1Q2 lattice ($k = 1, 2$). Even though MT-0 assumes that such temperature depends only on time, a 1-D approach is adopted so that $f_k = f_k(x,t)$, which it is indeed the case of MT-1. As heat is transferred by conduction within a stationary medium (i.e., catering meal), Eq. (12) suggests the following equilibrium distribution function:

$$f_k^{\text{eq}}(x,t) = w_k\, T(x,t) \quad , \quad w_1 = w_2 = 1/2 \quad (60)$$

In line with Eq. (17), relaxation factors $\omega_{\text{MT-0}}$ and $\omega_{\text{MT-1}}$ are different for each thermal model:

$$\text{MT-0: } \omega_{\text{MT-0}} = 2\text{, MT-1: } \frac{1}{\omega_{\text{MT-1}}} = \frac{\alpha_{\text{meal}}\Delta t}{(\Delta x)^2} + \frac{1}{2} \quad (61)$$

Since f_0 is disregarded in D1Q2, meal temperature is retrieved from distribution functions as:

$$T(x,t) = f_1(x,t) + f_2(x,t) \quad (62)$$

Table 3. Representative values for thermophysical parameters of catering meals

Thermophysical parameter	Representative value
Horizontal area for heat transfer	$A_{\text{inf}} = A_{\text{sup}} = A = 0.17225$ m^2
Meal density	$\rho_{\text{meal}} = 1014.4$ kg/m^3
Meal height within container	$L_{\text{meal}} = 0.030$ m ($L_{1/2} = L_{\text{meal}}/2$)
Meal mass within container	$m_{\text{meal}} = \rho_{\text{meal}}\, L_{\text{meal}}\, A = 5.242$ kg
Meal specific heat capacity	$c_{\text{p,meal}} = 3545.5$ J/(kg·K)
Meal thermal conductivity	$\lambda_{\text{meal}} = 0.488$ W/(m·K)
Meal thermal diffusivity	$\alpha_{\text{meal}} = 1.357 \times 10^{-7}$ m^2/s

Bearing in mind Eq. (55) and (58), one must impose $T(x,0) = T_0$ as the initial condition for both MT-0 and MT-1, which leads to:

$$f_1(x,0) = w_1 T_0 \quad , \quad f_2(x,0) = w_2 T_0 \tag{63}$$

Mixed-typed boundary conditions for MT-1, Eq. (59), are implemented in LBM by "mixing" the numerical procedures for Dirichlet condition, as in Eq. (36), and (this time non-null) Neumann condition, as in Eq. (37). Not required for MT-0, boundary conditions are compensated by the inclusion of a sink term in the collision step for MT-0, as in Eq. (15).

Based on TM-0, LBM simulations were first performed for blast-cooling of a meal sample initially at T_0 = 66.1°C, with air at allegedly constant temperature T_∞ = −13°C. As Figure 4(*a*) shows, LBM profiles simulated either with no streaming step or periodic boundary conditions are in good agreement with the analytical solution, Eq. (56). LBM profile with periodic boundary conditions is the one at midpoint site. Despite not shown for clarity, exactly the same profile was obtained for all sites in the D1Q2 lattice, which is consistent with a basic hypothesis of lumped-parameter analysis, namely, uniform temperature within the system.

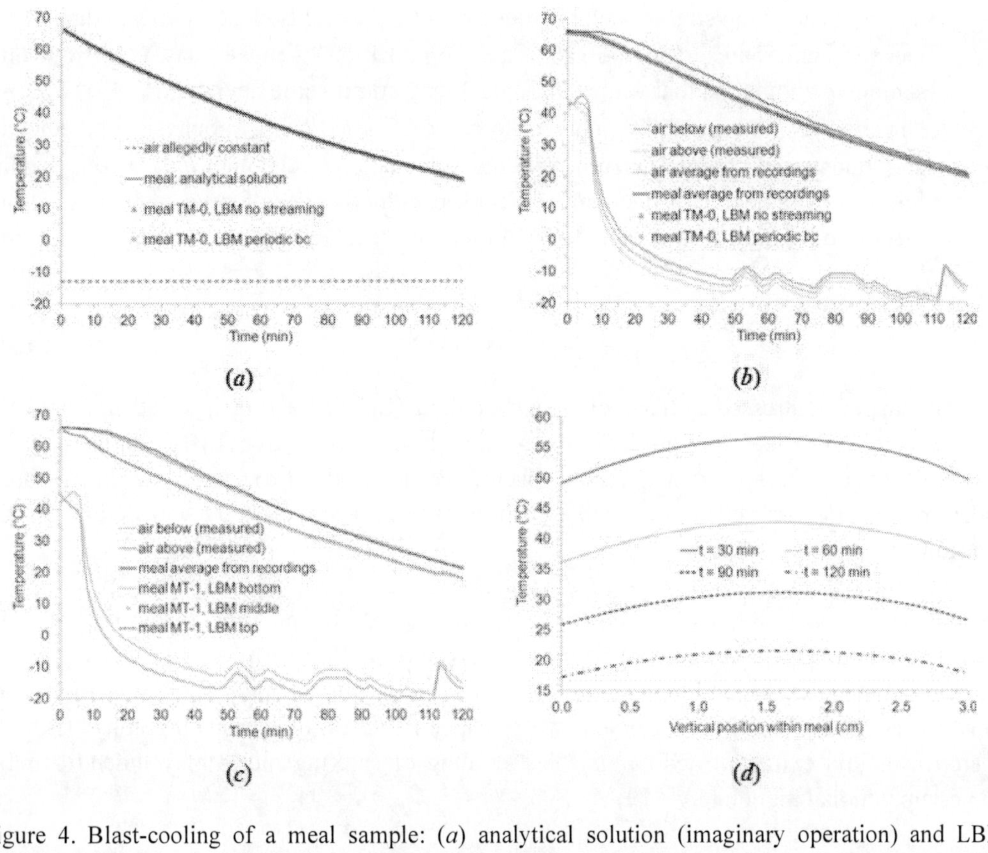

Figure 4. Blast-cooling of a meal sample: (*a*) analytical solution (imaginary operation) and LBM simulations using TM-0, (*b*) recorded profiles and LBM simulations using TM-0, (*c*) recorded profiles and LBM simulations using TM-1, (*d*) LBM simulations using TM-1 of temperature distribution within the meal sample at distinct instants during the operation.

For further comparisons, LBM simulations based on TM-0 were carried out using profile $T_{air,av}(t)$ obtained via Eq. (53) from profiles $T_{air,inf}(t)$ and $T_{air,sup}(t)$ actually recorded for cooling air. Together with such air profiles, Figure 4(*b*) shows LBM simulations and average meal profile assessed from recordings for meat and sauce. Overall behavior is reasonably reproduced under both approaches for the streaming step and good agreement occurs at the operation end. Peaks in air profiles are due to casual openings of the blast-cooling cell door for inspection purposes.

LBM simulations based on TM-1 were next performed using recorded profiles $T_{air,inf}(t)$ and $T_{air,sup}(t)$ for the corresponding mixed-type boundary conditions, Eq. (59). One should note that $T_{air,av}(t)$ is no longer necessary. As shown in Figure 4(*c*), good agreement now occurs throughout the operation, under both approaches for the streaming step. For inclusiveness, profiles simulated at meal bottom ($x = 0$) and top ($x = L_{meal}$) are also shown and they are cooler, as expected. Figure (*d*) presents the spatial distribution of temperature over the meal height as simulated via LBM based on TM-1 after 30, 60, 90 and 120 minutes. Lower profiles are obtained inasmuch as blast-cooling operation proceeds, likewise expected.

Last but not least, the following points are worth pondering:

- LBM simulations are compared to profiles supposedly recorded at meal core (midpoint), but recorders may have undesirably moved inside the food container during its handling so that such profiles may refer to points closer to meal edge;
- Despite comprising meat and sauce, the composition of meals may vary from one sample to another so that values in Table 3 may admit some deviation;
- Uncertainties and variability inherent to recordings and/or thermophysical parameters are "transferred" to coefficient h_c via model equations. Although $h_{c,inf} \neq h_{c,sup}$ would bring an extra degree of freedom, vertical separation between food containers suggests that those coefficients are likely to be of the same order, as discussed in [50].

CONCLUSION

This chapter addressed lattice-Boltzmann method (LBM) bearing in mind not only its basic concepts but also the proper implementation of prospective LBM simulators for agroindustrial processes. For that reason, discussions focused on dynamic one-dimensional model frameworks as well as on heat and mass (species) transport by convection and/or diffusion.

Yet, even when such simplifying assumptions are invoked, transport phenomena involving agroindustrial biosystems are likely to require the numerical solution of stiff governing equations. LBM comes forward as alternative and powerful tool to study process behavior under distinct and non-tested conditions. As this chapter suggests, food and bioprocesses like (but not restricted to) biospecific affinity chromatography (BAC), supercritical fluid extraction (SFE) and blast-cooling of catering meals may indeed benefit from computational modeling via LBM.

REFERENCES

[1] Romano, V. R. *J. Food Eng.* 2005, 71, 231-232.
[2] Datta, A. K.; Sablani, S. S. In: *Handbook of Food and Bioprocess Modeling Techniques*; Sablani, S. S.; Rahman, M. S.; Datta, A. K.; Mujumdar, A. R. (Eds.); CRC Press: Boca Raton, 2007; pp. 1-11.
[3] Wang, L.; Sun, D. W. *Trends Food Sci. Tech.* 2003, 14, 408-423.
[4] Sun, D. W. *Comp. Elect. Agric.* 2002, 34, 1-3.
[5] Norton, T.; Sun, D. W. In: *Computational Fluid Dynamics in Food Processing*; Sun, D. W. (Ed.).; CRC Press: Boca Raton, 2007, pp. 1-41.
[6] Migliori, M.; Gabriele, D.; de Cindio, B.; Pollini, C. M. *J. Food Eng.* 2005, 71, 242-251.
[7] Erdogdu, F.; Sarkar, A.; Singh, R. P. *J. Food Eng.* 2005, 71, 287-294.
[8] Puri, V. M.; Anantheswaran, R. C. *J. Food Eng.* 1993, 19, 247-274.
[9] Romano, V. R.; Marra, F.; Tammaro, U. *J. Food Eng.* 2005, 71, 233-241.
[10] Curcio, S.; Calabro, V.; Iorio, G. *J. Food Eng.* 2005, 71, 252-264.
[11] Yu, C.; Gunasekaran, S. *J. Food Eng.* 2005, 71, 295-303.
[12] Scott, G.; Richardson, P. *Trends Food Sci. Tech.* 1997, 8, 119-124.
[13] Xia, Bi; Sun, D. W. *Comp. Elect. Agric.* 2002, 34, 5-24.
[14] van der Sman, R. G. M. *J. Food Eng.* 2003, 60, 383-390.
[15] Mohamad, A. A. *Lattice Boltzmann Method: Fundamentals and Engineering Applications with Computer Codes*; Springer-Verlag: London, 2011.
[16] Rothman, D. H.; Zaleski, S. *Lattice-gas cellular automata: simple models of complex hydrodynamics*; Cambridge University Press: Cambridge, 1997.
[17] Wolf-Gladrow, D. A. *Lattice-Gas Cellular Automata and Lattice Boltzmann Models: an Introduction*; Springer: Berlin, 2000.
[18] Succi, S. *The Lattice Boltzmann Equation for Fluid Dynamics and Beyond*; Oxford University Press: New York, 2001.
[19] van der Sman, R. G. M. In: *Handbook of Food and Bioprocess Modeling Techniques*; Sablani, S. S.; Rahman, M. S.; Datta, A. K.; Mujumdar, A. R. (Eds.); CRC Press: Boca Raton, 2007; pp. 15-39.
[20] van Kats, F. M.; Egberts, P. J. P. *Transp. Porous Media.* 1999, 37, 55-68.
[21] van der Sman, R. G. M. (1999). *Lattice Boltzmann schemes for convection-diffusion phenomena - application to packages of agricultural products* [Ph.D. Thesis]. Available at: <http://library.wur.nl/WebQuery/clc/963315>, accessed in November 2011.
[22] Mantle, M. D.; Bijeljic, B.; Sederman, A. J.; Gladden, L. F. *Magnet. Reson. Imag.* 2001, 19, 527-529.
[23] Ladd, A. J. C.; Verbeg, R. *J. Stat. Phys.* 2001, 104, 1191-1251.
[24] Feng, Z. G.; Michaelides, E. E. *J. Hydr. Eng.* 2003, 129, 985-994.
[25] Zhang, X. X.; Bengough, A. G.; Deeks, L. K.; Crawford, J. W.; Young, I. M. *Water Res. Res.* 2002, 38, 1167-1176.
[26] Zhang, X. X.; Ren, L. *J. Contam. Hydr.* 2003, 67, 27-42.
[27] Migliorini, C.; Qian, Y.; Chen, H.; Brown, E. B.; Jain, R. K.; Munn, L. L. *Biophys. J.* 2002, 83, 1834-1841.
[28] Sun, C.; Migliorini, C.; Munn, L. L. *Biophys. J.* 2003, 85, 208-222.

[29] Santos, L. O. E.; Philippi, P. C.; Damiani, M. C.; Fernandes, C. P. *J. Petr. Sci. Eng.* 2002, 35, 109-124.
[30] Santos, L. O. E.; Facin, P. C.; Philippi, P. C. *Phys. Rev. E.* 2003, 68, 056302 (10 pages).
[31] Facin, P. C.; Philippi, P. C.; Santos, L. O. E. *Fut. Gen. Comp. Sys.* 2004, 20, 945-949.
[32] Sukop, M. C.; Or, D. *Water Res. Res.* 2004, 40, W01509, doi:10.1029/2003WR002333.
[33] Schure, M. R.; Maier, R .S.; Kroll, D. M.; Davis, H. T. *J. Chrom. A.* 2004, 1031, 79-86.
[34] Surmas, R.; Santos, L. O. E.; Philippi, P. C. *Fut. Gen. Comp. Sys.* 2004, 20, 951-958.
[35] Qian, Y. H.; D'Humières, D.; Lallemand, P. *Europhys. Lett.*, 1992, 17, 479-484.
[36] Maier, R. S.; Bernard, R. S.; Daryl, W. G. *Phys. Fluids*, 1996, 8, 1788-1801.
[37] Karlin, I. V.; Ansumali, S.; Frouzakis, C. E.; Chikatamarla, S. S. *Commun. Comput. Phys.*, 2006, 1, 1-45.
[38] Kamimura, E. S.; Mediata, O.; Rodrigues, M. I.; Maugeri Filho, F. *Biotechnol. Applied Biochem.*, 2001, 33, 153-159.
[39] Chase, H. A. *J. Chromatogr.*, 1984, 297, 179-202.
[40] Della Porta, G.; Taddeo, R.; D'Urso, E.; Reverchon, E. *LWT – Food Sci. Technol.*, 1998, 31, 454-460.
[41] Subra, P.; Castellani, S.; Jestin, P.; Aoufi, A. *J. Supercrit. Fluids*, 1998, 12, 99-104.
[42] Oliveira, A. L.; Silva, S. S.; da Silva, M. A. P.; Cabral, F. A. *J. Food Sci. Technol.*, 2001, 38, 38-42.
[43] Doyle, E. (2002). *Survival and growth of Clostridium perfringens during the cooling step of thermal processing of meat products* [FRI Briefings]. Available at: <http://www.wisc.edu/fri/briefs/cperfsurvivgrow.pdf>, accessed in June 2009.
[44] Crouch, E.; Golden, N. J. (2005). *A risk assessment for Clostridium perfringens in ready-to-eat and partially cooked meat and poultry products* [Report]. Available at: <http://www.fsis.usda.gov/PDF/CPerfringens_Risk_Assess_Sep2005.pdf>, accessed in June 2009.
[45] Bellara, S. R.; McFarlane, C. M.; Thomas, C. R.; Fryer, P. J. *Chem. Eng. Sci.*, 2000, 55, 6085-6095.
[46] Amézquita, A.; Weller, C. L.; Wang, L.; Thippareddi, H.; Burson, D. E. *Int. J. Food Microbiol.*, 2005, 101, 123-144.
[47] Almonacid, S.; Simpson, R. Teixeira, A. *J. Food Sci.*, 2007, 72, E508-E517.
[48] Nield, D. A.; Bejan, A. *Convection in Porous Medium*. Springer: New York, 1999.
[49] Rabi, J. A.; Mohamad, A. A. In: *Discrete Simulation of Fluids Dynamics – DSFD 2008*; UFSC: Florianópolis, 2008; p. 25.
[50] Rabi, J. A.; Trezzani-Harbelot, I.; Morelli, E.; Guilpart, J. *Int. Rev. Mech. Eng.*, 2012, 6, 218-227.
[51] Jaloustre, S.; Cornu, M.; Morelli, E.; Noël, V.; Delignette-Muller, M. L. *Food Microbiol.*, 2011, 28, 311-320.
[52] Kays, W. M.; Crawford, M. E. *Convective Heat and Mass Transfer*. McGraw-Hill: New York, 1993.
[53] Ching, C. B.; Wu, Y. X.; Lisso, M.; Wozny, G.; Laiblin, T.; Arlt, W. *J. Chromatogr. A*, 2002, 945, 117-131.
[54] Özdural, A. R.; Alkan, A.; Kerkhof, P. J. A. M. *J. Chromatogr. A*, 2004, 1041, 77-85.
[55] Leickt, L.; Månsson, A.; Ohlson, S. *Anal. Biochem.*, 2001, 291, 102-108.
[56] Cowan, G. H.; Gosling, I. S.; Laws, J. F.; Sweetenham, W. P. *J. Chromatogr.*, 1986, 363, 37-56.

[57] Sridhar, P.; Sastri, N. V. S.; Modak, J. M.; Mukherjee, A. K. *Chem. Eng. Technol.*, 1994, 17, 422-429.

[58] Kempe, H.; Axelsson, A.; Nilsson, B.; Zacchi G. *J. Chromatogr. A*, 1999, 846, 1-12.

[59] Yun, J.; Lin, D.-Q.; Yao, S.-J. *J. Chromatogr. A*, 2005, 1095, 16-26.

[60] Patankar, S. V. *Numerical Heat Transfer and Fluid Flow*. Hemisphere: New York, 1980.

[61] Ferziger, J. H., Perić, M. *Computational Methods for Fluid Dynamics*. Springer-Verlag: Berlin, 2002.

[62] Campos, L. M. A. S.; Michielin, E. M. Z.; Danielski, L.; Ferreira, S. R. S. *J. Supercri. Fluids*, 2005, 34, 163-170.

[63] Sovová, H. *Chem. Eng. Sci.*, 1994, 49, 409-414.

[64] Reverchon, E. *AIChE J.*, 1996, 42, 1765-1771.

[65] França, L. F.; Meireles, M. A. A. *J. Supercrit. Fluids*, 2000, 18, 35-47.

[66] Wu, W.; Hou, Y. *J. Supercrit. Fluids*, 2001, 19, 149-159.

[67] Lucas, S.; Calvo, M. P.; García-Serna, J.; Palencia, C.; Cocero, M. J. *J. Supercrit. Fluids*, 2007, 41, 257-266.

[68] Gaspar, F.; Lu, T.; Santos, R.; Al-Duri, B. *J. Supercrit. Fluids*, 2003, 25, 247-260.

[69] Vargas, R. M. F.; Cassel, E.; Gomes, G. M. F.; Longhi, L. G. S.; Atti-Serafini, L.; Atti-Santos, A. C. *Braz. J. Chem. Eng.*, 2006, 23, 375-382.

[70] Özişik, M. N. *Heat Transfer: A Basic Approach*. McGraw-Hill: New York, 1985.

[71] van der Sman, R. G. M. *J. Food Eng.*, 2003, 60, 383-390.

In: Advances in Computational Modeling Research
Editor: Anna Belya Kora
ISBN: 978-1-62618-065-9

Chapter 5

AGENT-BASED MODELS IN CANCER PREVENTION RESEARCH

John W. Pepper, Nadarajen A. Vydelingum, Barbara K. Dunn and Richard M. Fagerstrom
Division of Cancer Prevention, National Cancer Institute, Bethesda, MD, US

ABSTRACT

Agent-based models have proven very useful for studying complex dynamic processes. As biomedical researchers have turned their attention to the most complex diseases, they have made increasing use of agent-based models in various areas, including cancer biology. Cancer is arguably the ultimate complex biological system. It has recently become clear that much of its complexity arises through Darwinian evolution, driven by uncontrolled natural selection among rapidly mutating somatic cells.

Here we review the application of agent-based models and simulations to five key open questions in cancer prevention research. The five research questions we review as case studies include: 1) What normally prevents somatic evolution from generating cancer? 2) Why do benign growths often progress to malignant cancers? 3) What is the chronological sequence of molecular events in cancer progression? 4) Are there reliable molecular biomarkers for cancer?; and 5) Will evolved drug resistance stymie efforts at long-term cancer chemoprevention?

We conclude that molecular analysis can be usefully augmented by using agent-based models to generate plausible and productive hypotheses for empirical testing.

Despite substantial progress in the US "war on cancer" declared in 1971, cancer incidence and mortality have fallen less than initially hoped and expected (Bailar and Smith 1986). Even within the most recent five years, cancer incidence in the US declined by less than 1%, and cancer mortality declined by less than 2% (Siegel, Naishadham et al. 2012). This has led some researchers to focus on the complexities of the disease, and to call for new approaches to cancer medicine that can explicitly address the complex dynamics that arise from the cellular evolution and adaption that occurs in cancer, as it interacts with its own micro-ecology (Gatenby 2009; Drake 2011).

Although the research strategy of using molecular reductionism to define a fixed disease-causing entity has a brilliant track record in infectious disease and inherited genetic disease, it has been less successful in cancer biology. Molecular analysis quickly revealed the consistent genetic nature of non-human pathogen organisms such as bacteria, and of inherited genetic defects. In contrast, cancer reflects a complex process of ongoing genetic change in human somatic cells, and it usually has no fixed molecular basis. Thus, in important ways, a cancer is not simply a *thing* to be removed or destroyed, but also a dynamic *process* to be controlled or prevented. This process is pathological Darwinian evolution among somatic cells.

REPRESENTING THE CENTRAL PROCESS OF CANCER BIOLOGY: SOMATIC CELLULAR EVOLUTION

Agent-based computer models have become important in cancer biology, because they are established as being as being effective for modeling and understanding dynamic processes such as adaptive evolution, e.g., (An 2010).

As early as the 1970's, Cairns (Cairns 1975) and Nowell (Nowell 1976) proposed that somatic cell mutation, selection, and resulting Darwinian evolution are the fundamental processes by which tumors arise, acquire malignancy, and evade therapy. This hypothesis lay dormant for decades before being further developed and confirmed with empirical results from modern cancer biology (Frank and Nowak 2004; Crespi and Summers 2005; Pepper, Findlay et al. 2009; Greaves and Maley 2012; Podlaha, Riester et al. 2012). Currently, "cancer development at the cellular level is widely regarded as a Darwinian evolutionary process involving natural selection of genetically variant cells in the context of a complex micro-environmental ecology" (Anderson, Lutz et al. 2011). Our only amendment to the foregoing quote would be to explicitly recognize that relevant mechanisms of cellular heredity include both genetic and epigenetic variation (Podlaha, Riester et al. 2012).

As molecular techniques have provided an increasing wealth of data, the hypothesis of cellular evolution in cancer has been supported in various organs. Somatic cellular evolution has been documented in molecular detail in breast cancer (Shah, Morin et al. 2009; Navin, Kendall et al. 2011; Shah, Roth et al. 2012), pancreatic cancer (Campbell, Yachida et al. 2010), renal cancer (Gerlinger, Rowan et al. 2012), ovarian cancer (Ng, Cooke et al. 2012), leukemia (Anderson, Lutz et al. 2011; Walter, Shen et al. 2012), colorectal cancer (Hirsch, Camps et al. 2012), and head and neck cancer (Masuda, Toh et al. 2012).

The hypotheses and observations of cellular evolution are central contributors to an emerging scientific theory of cancer dynamics, which elucidates origination, progression, metastasis, and response to treatment. This enterprise is greatly assisted by a well-developed body of concepts and techniques from the classical study of Darwinian evolution in populations of organisms. One optimistic assessment is that, despite the extreme complexity of cancer, "most, if not all, of this complexity can be explained by classical evolutionary principles" (Greaves and Maley 2012).

Agent-based Models for Studying Evolutionary Processes

Armed with an understanding of the basic process behind the complex dynamics of cancer, researchers are increasingly adopting agent-based computer models, which have proven effective for understanding complex dynamic processes such as cellular evolution. This technique allows researchers to reveal not only the molecular components of cancer cells, but also the reasons why molecular components vary among cells, and why they change over time during cancer progression. Cancer is arguably the ultimate complex biological system (Hatzikirou, Chauviere et al. 2012). For coping with such complexity, systems biology has gained popularity in much of biomedicine, including cancer biology (Rosenfeld and Kapetanovic 2008; Baker and Kramer 2011; Mijušković, Brown et al. 2012). Systems biology includes both methods for analyzing large 'omics' data sets, and also methods, such as agent-based models, for understanding the processes that generate those data.

Agent-based models and simulations can generate a process of 'digital evolution' inside a computer, which can provide a detailed, transparent, and predictive model of cellular evolution and progression under a specified set of conditions (Adami and Wilke 2004; Pennock 2007). The results of such simulations can generate incisive new hypotheses that are empirically testable, and that can guide empirical research into productive new directions. The loop between experiments and models is thereby closed, allowing continued iterations as new empirical results lead to improved theory and new testable hypotheses (Krakauer, Collins et al. 2011).

When applied to cancer biology, agent-based models usually represent a population of tumor cells, with each cell represented by its own software 'agent' that can interact with its environment, live or die, and divide into two daughter cells that inherit the traits of the parent cell. Each time step of such a model is typically calibrated to represent the generation time of the cells. Over multiple cell generations, the same kind of Darwinian dynamics that actually drive change in real cancer cell populations also arise in this virtual process of 'digital evolution', where they are more easily observed, measured, and experimented upon. The crucial advantage is that in the virtual world, we can measure any variable, and even manipulate any variable, making it possible to explore an unlimited range of hypotheses for logical consistency, and for consistency with empirical observations.

Agent-based Models in Cancer Prevention Research

The process of cellular evolution inside the body can rarely be observed directly, and most empirical data can only represent a 'snapshot' in time from this dynamic process. Explicitly representing the process that is believed to underlie such observable data extends their explanatory power. Below, we consider five current questions in cancer prevention research, and discuss the application of agent-based models to each of them.

What Normally Prevents Somatic Evolution from Generating Cancer?

Once it is clear that cancer arises from natural selection among somatic cells, an evolutionary perspective immediately raises the question of how this process is normally avoided, so that most people never get cancer. Darwinian evolution among somatic cells, if allowed to proceed unrestrained, would inevitably remove all normal limitations on cell survival and proliferation, because these reduce the Darwinian fitness of cells. Removal of all normal limitations on cell survival and proliferation is a good cell-level description of cancer. From this perspective, the question of why some people get cancer may seem less compelling than the question of how anyone avoids cancer through an entire life time. A proposed answer to this question was offered as a hypothesis that the characteristic cellular structure of most tissues in most animals has the precise function of preventing the inheritance of cellular traits between successive generations of mature cells within tissues. (Pepper, Sprouffske et al. 2007). Because inheritance of traits from one generation to the next is necessary for Darwinian evolution, this simple expedient, according to these authors, is sufficient to largely suppress somatic evolution, thereby relegating cancer to being a disease only of advanced age, after a very long period for weak somatic evolution. To evaluate the plausibility of this novel hypothesis, the authors used an agent-based model to conduct simulation experiments on a population of virtual somatic cells inside a simulated tissue. This 'thought experiment' made it possible to manipulate fundamental properties of cells and tissues that could not be manipulated in a living preparation. The results were in accord with theoretical predictions (Figure 1), lending the hypothesis enough plausibility to be considered worthy of empirical testing.

Why Do Benign Neoplasms often Progress to Malignancy?

The defining (and lethal) feature of cancer, as opposed to benign growths, is tumors that invade surrounding tissues, ultimately leading to distant metastasis. Because this process emerges from interactions among multiple cells and their micro-environments, agent-based models are particularly well suited to its study. For example, ecologists have used discrete numerical simulations of spatiotemporal dynamics to compare tissue invasion by cancer cells with the invasion of natural ecosystems by artificially introduced species (Marco, Cannas et al. 2009). These authors posit that their results can help explain the patterns of genetic diversity observed both within and among tumors in a patient. They also describe a "distinct geometrical signature" of invasion dynamics, which they suggest may someday be used along with advanced imaging technologies to accurately discriminate malignancies from benign growths (Martin 2008).

A central goal of cancer prevention is to block progression from benign pre-malignancies to invasive cancers. At a cellular level, this progression entails the acquisition of motility by formerly sedentary tumor cells. In seeking ways to block this change, it will help to know how it occurs. Building on classical evolutionary ecology, one research group (Aktipis, Maley et al. 2012) hypothesized that the intense local competition for vital resources caused by the abnormally rapid cell proliferation in tumors selects for motile cells that can thrive by leaving the resource-depleted zones where stationary tumor cells languish and die, and that this selection drives the evolution of tumor cell motility. Simulations using agent-based computational models provided preliminary support for this hypothesis. Their results suggest that therapeutic drugs designed to block cell motility, rather than to kill tumor cells, can block malignancy and also provide other important advantages. Cells blocked from evolving motility will also be selected to adapt to their impoverished environment by reducing cell proliferation and thus tumor growth. In contrast to the evolutionary response to cytotoxic drugs designed to kill cancer cells, motility-blocking drugs are predicted to be more robust against acquired drug resistance, which causes many cancer therapies to lose effectiveness before the patient is cured (Aktipis, Maley et al. 2012).

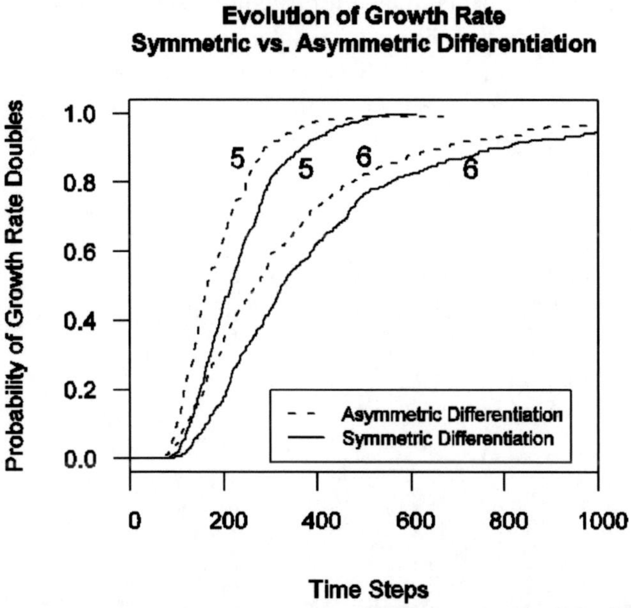

Figure 1. Results of agent-based simulations of the evolution of somatic cell hyper-proliferation. Horizontal axis shows time, and vertical axis shows probability of doubling the population average growth rate. Inheritance between generations of mature functional cells is prevented when lost cells are replaced only through the division of sequestered somatic stem cells, and the symmetric division of transient amplifying cells that do not self-renew their own population. Under this condition, hyper-proliferation is substantially delayed ($p < 0.001$). The numbers 5 and 6 refer to the number of transient amplifying stages between tissue stem cells and mature functional cells. With more amplifying stages, the population of stem cells is smaller, presenting less potential for somatic evolution. After (Pepper, Sprouffske et al. 2007).

What Is the Chronological Sequence of Molecular Events in Cancer Progression?

A long-standing challenge in cancer biology is to understand and predict the chronological sequence of molecular events underlying the initiation and progression of tumors (Fearon and Vogelstein 1990). Many early genetic models assumed that tumors were genetically homogenous, and thus could be characterized as having a single tumor genotype. On this basis, the classic type of model was a linear sequence of events, or a ‘canonical path model’, constructed from cross-sectional data on multiple patients (Sprouffske, Pepper et al. 2011). However, more recent evidence has revealed so much genetic heterogeneity within tumors that the assumption of one single genotype per tumor is clearly no longer valid (Nord 2011; Shibata 2012).

Agent-based evolutionary models of tumor progression suggest that the temporal, or evolutionary, order of mutations acquired in the clones that survive over the lifetime of a neoplasm may not be consistent with the path order inferred from cross-sectional data (Sprouffske, Pepper et al. 2011). Because cellular evolution during tumorigenesis is a unique process in each patient, it is not valid to assume that the state of one tumor is informative for the history of even an apparently similar tumor in a different patient. Independent evolutionary trajectories in each patient make it difficult to reconstruct temporal order from cross-sectional data. In contrast, observed cellular diversity can be analyzed using phylogenetic methods from classical evolutionary biology to reconstruct cell lineages within individual tumors. This can accurately reveal the true temporal order of events for the clones within a specific tumor (Figure 2).

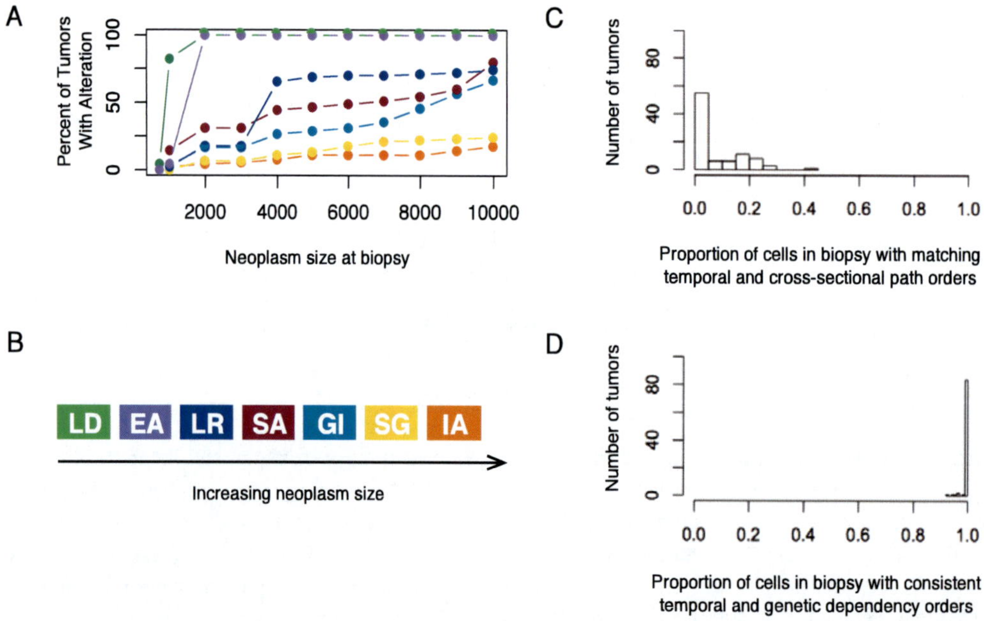

Figure 2. Reprinted with permission from (Sprouffske, Pepper et al. 2011): The temporal order of mutations seen in simulated cancer clones rarely matches the path order inferred from cross-sectional data, but it does match the order inferred from genetic dependency analysis of intratumor data. (A): Plotting the percentage of tumors with a given mutation at increasing neoplasm sizes can be used to infer (B): the cross-sectional path model of mutations. (C): However, the proportion of cells within a simulated neoplasm that have a history consistent with the cross-sectional path order tends to be low (mean [SEM] = 7.3% [1.0%], n = 90). (D): In contrast, the proportion of cells within a simulated neoplasm that have a history consistent with order inferred from the genetic dependency analysis is high (mean [SEM] = 99.7% [0.1%], n = 90). Each mutation type is represented by a different color in panels A & B: loss of differentiation (LD) is green, evasion of apoptosis (EA) is light purple, limitless replicative potential (LR) is orange, sustained angiogenesis (SA) is dark purple, genomic instability (GI) is yellow, self-sufficiency in growth signals (SG) is light blue, and insensitivity to antigrowth signals (IA) is dark blue.

The two key results of this study (Sprouffske, Pepper et al. 2011) are: (1) that cross-sectional data can be misleading because of independent cellular evolution within each patient; and (2) that even though detailed study of intratumor variation can reveal the average ordering of molecular events within that tumor, not all cells in the tumor actually reflect that average order of events. As with any results from rigorous simulations, these should not be treated as final conclusions, but rather as plausible hypotheses and strong candidates for empirical testing.

Are There Reliable Molecular Biomarkers for Cancer?

Another major research direction in cancer prevention is the attempt to identify molecules that can serve as 'cancer biomarkers' for early detection. The hope here is that molecules can be discovered that are characteristic of cancer, but not of normal tissue. To date, most biomarker development has not been informed by evolutionary dynamics, but instead has focused on examining specific molecules one by one. This strategy has identified many candidate somatic genetic and epigenetic 'biomarkers' of cancer or cancer risk. However, very few of these candidate molecules have proven to be reliably effective for identifying cancer (Ransohoff 2009). Similarly, no specific molecule has yet been validated as a robust predictor of progression to cancer, or been useful to reduce cancer mortality (Li, Blount et al. 2011).

Will Evolved Drug Resistance Stymie Efforts at Long-term Cancer Chemoprevention?

Developing drugs for chemoprevention is a major line of research in cancer prevention that may face pitfalls from cellular evolution. These potential pitfalls have already been demonstrated in cancer therapy. Many agents have been developed and used successfully for killing cancer cells. A very common limitation on this approach, however, is the acquired drug resistance that results from cellular evolution (Pepper 2011). When a tumor acquires drug resistance, this often leads to relapse rather than cure. (Cunningham, Gatenby et al. 2011; Pepper 2011), and most patients who die of cancer are killed by a cancer that has evolved drug resistance. The problem of acquired resistance may prove to be even more limiting for long-term chemoprevention than it is for therapy. While successful therapy is completed within a shorter time frame, cancer prevention might ideally continue throughout life. Thus, any prevention drug that can be administered effectively only for a limited time may not be entirely satisfactory.

There is clinical evidence that the time limits on effective drug administration seen during cancer therapy can also arise in chemoprevention.

Achieving life-long protection through chemoprevention may require longer periods of effective drug administration, and thus not be possible with any drug that selects for drug resistant variants. Whenever somatic cells are dividing and subject to mutation, resistant variants are likely to arise. Any drug that kills or suppresses individual drug-sensitive cells thereby 'selects' more resistant cells to survive with reduced competition after the removal of drug-sensitive neighbors (Pepper 2008). This leads to rapid out-growth, or 'clonal expansion', of a lineage of resistant cells which quickly takes the place of the drug-sensitive cells removed by treatment.

Providing the best possible protection through chemoprevention may require ways to reduce cellular evolution of acquired resistance by reducing Darwinian selection for resistance among cells. Classical evolutionary theory suggests specific strategies (Pepper 2008), which have been developed and validated using agent-based models of the evolution of resistance against potential new classes of drugs. Therapeutics that change the tumor micro-environment to make it less hospitable to tumor growth and invasion can impede all cancer cells equally, without selecting among them and thereby driving the evolution of resistance. Such existing and potential drug classes could exert less selection on existing variation, and thus slow the evolution of acquired drug resistance. In contrast to most cancer drugs, which are designed to kill cancer cells, potential new drugs could exploit the fact that cancer cells construct their own micro-environment by producing, and depending on, many shared substances that improve the capacity of their shared micro-environment to support tumor growth and invasion. Examples include secreted angiogenesis factors, growth and invasion factors, and immune suppression factors (Pepper 2008; Pepper 2011). In agent-based computer simulations, drugs targeting such shared 'public goods' products do not single out the specific cells producing them, or those cells producing drug-resistant variants of them. Thus they do not select strongly for resistant cells, and do not strongly drive the evolution of acquired resistance (Pepper 2012) (Figure 3). This raises the hope that such drugs would retain longer-term effectiveness both in treatment, and also in chemoprevention. Specific targets for potential drugs of this class have been proposed, and in some cases successfully tested in animal models (Pepper 2008; Pepper 2011).

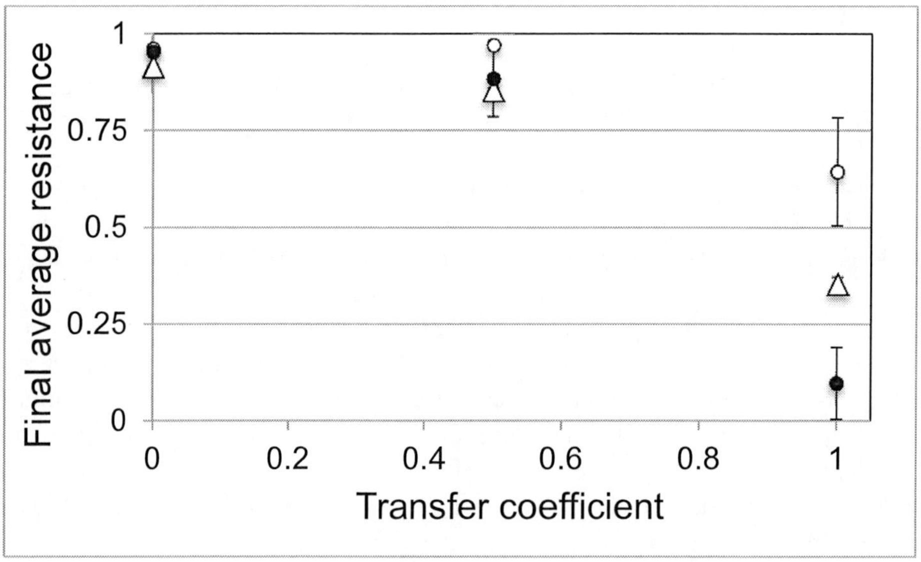

Figure 3. Simulation results reprinted with permission from (Pepper 2012): Frequency of drug resistant cells after 1000 cell generations of mutation and evolution. High transfer coefficients correspond to highly diffusible drug targets produced by one cell that provide benefits to many other neighboring cells. A transfer coefficient of zero corresponds to a cell-intrinsic target molecule that is not shared among neighboring cells. Each marker represents the mean for a different drug dosage: Filled circle = 100% of full dose, open circle = 75%, triangle = 50%. Error bars show standard error across 10 simulation runs with different seed values for the pseudorandom number generator.

Conclusion

Viewing cancer simply as an entity to be removed or destroyed can overlook important aspects of cancer biology, including why cancer is dynamic and adaptive, and why it's difficult to target as a 'non-self' entity. Molecular reductionism contributes greatly to advances in cancer research, but is even more useful within the framework of understanding the somatic cellular evolution that generates the diverse molecules observed in cancer cells.

At the heart of cancer biology is change over time in populations of initially normal human cells. This process is best understood as somatic cellular evolution. Agent-based models are useful tools for understanding this process to generate the most crucial hypotheses for empirical testing.

Acknowledgments

For helpful input, we thank Carlo C. Maley, and Brian J. Reid.

References

Adami, C. and C. O. Wilke (2004). "Experiments in digital evolution (editors' introduction to the special issue)." *Artificial Life* 10(2): 117-122.

Aktipis, C. A., C. C. Maley, et al. (2012). "Dispersal Evolution in Neoplasms: The Role of Disregulated Metabolism in the Evolution of Cell Motility." *Cancer Prevention Research* 5(2): 266-275.

An, G. C. (2010). "Translational systems biology using an agent-based approach for dynamic knowledge representation: An evolutionary paradigm for biomedical research." *Wound Repair and Regeneration* 18(1): 8-12.

Anderson, K., C. Lutz, et al. (2011). "Genetic variegation of clonal architecture and propagating cells in leukaemia." *Nature* 469(7330): 356-+.

Bailar, J. C. and E. M. Smith (1986). "Progress against Cancer?" *New England Journal of Medicine* 314(19): 1226-1232.

Baker, S. G. and B. S. Kramer (2011). "Systems biology and cancer: Promises and perils." *Progress in Biophysics & Molecular Biology* 106(2): 410-413.

Cairns, J. (1975). "Mutation Selection and the Natural History Of Cancer." *Nature* 255(5505): 197-200.

Campbell, P. J., S. Yachida, et al. (2010). "The patterns and dynamics of genomic instability in metastatic pancreatic cancer." *Nature* 467(7319): 1109-1113.

Crespi, B. and K. Summers (2005). "Evolutionary biology of cancer." *Trends in Ecology & Evolution* 20(10): 545-552.

Cunningham, J. J., R. A. Gatenby, et al. (2011). "Evolutionary dynamics in cancer therapy." *Molecular Pharmaceutics* 8(6): 2094-2100.

Drake, N. (2011). "Forty years on from Nixon's war, cancer research 'evolves'." *Nat Med* 17(7): 757-757.

Fearon, E. R. and B. Vogelstein (1990). "A Genetic model for colorectal tumorigenesis." *Cell* 61(5): 759-767.

Frank, S. A. and M. A. Nowak (2004). "Problems of somatic mutation and cancer." *Bioessays* 26(3): 291-299.

Gatenby, R. A. (2009). "A change of strategy in the war on cancer." *Nature* 459(7246): 508-509.

Gerlinger, M., A. J. Rowan, et al. (2012). "Intratumor Heterogeneity and Branched Evolution Revealed by Multiregion Sequencing." *New England Journal of Medicine* 366(10): 883-892.

Greaves, M. and C. C. Maley (2012). "Clonal evolution in cancer." *Nature* 481(7381): 306-313.

Hatzikirou, H., A. Chauviere, et al. (2012). "Integrative physical oncology." *Wiley Interdisciplinary Reviews-Systems Biology and Medicine* 4(1): 1-14.

Hirsch, D., J. Camps, et al. (2012). "A new whole genome amplification method for studying clonal evolution patterns in malignant colorectal polyps." *Genes Chromosomes & Cancer* 51(5): 490-500.

Krakauer, D. C., J. P. Collins, et al. (2011). "The challenges and scope of theoretical biology." *Journal of Theoretical Biology* 276(1): 269-276.

Li, X. H., P. L. Blount, et al. (2011). "Application of Biomarkers in Cancer Risk Management: Evaluation from Stochastic Clonal Evolutionary and Dynamic System Optimization Points of View." *Plos Computational Biology* 7(2).

Marco, D. E., S. A. Cannas, et al. (2009). "Comparable ecological dynamics underlie early cancer invasion and species dispersal, involving self-organizing processes." *Journal of Theoretical Biology* 256(1): 65-75.

Martin, M. (2008). "Comparing Invasive Species to Metastatic Cancers Inspires New Insights for Modelers." *Journal of the National Cancer Institute* 100(2): 88-89.

Masuda, M., S. Toh, et al. (2012). "Somatic evolution of head and neck cancer – Biological robustness and latent vulnerability." *Molecular Oncology*(0).

Mijušković, M., S. M. Brown, et al. (2012). "A Streamlined Method for Detecting Structural Variants in Cancer Genomes by Short Read Paired-End Sequencing." *PLoS ONE* 7(10): e48314.

Navin, N., J. Kendall, et al. (2011). "Tumour evolution inferred by single-cell sequencing." *Nature* 472(7341): 90-U119.

Ng, C. K., S. L. Cooke, et al. (2012). "The role of tandem duplicator phenotype in tumour evolution in high-grade serous ovarian cancer." *The Journal of pathology* 226(5): 703-712.

Nord, D. G., Ed. (2011). *Intratumor Diversity and Clonal Evolution in Cancer* Advances in Cancer Research. Oxford, Academic Press.

Nowell, P. C. (1976). "Clonal Evolution of Tumor-Cell Populations." *Science* 194(4260): 23-28.

Pennock, R. T. (2007). "Models, simulations, instantiations, and evidence: the case of digital evolution." *Journal of Experimental & Theoretical Artificial Intelligence* 19(1): 29-42.

Pepper, J. W., K. Sprouffske, et al. (2007). "Animal cell differentiation patterns suppress somatic evolution." *Plos Computational Biology* 3(12): 2532-2545.

Pepper, J. W. (2008). "Defeating Pathogen Drug Resistance : Guidance from Evolutionary Theory " *Evolution* 62(12): 3185-3191.

Pepper, J. W. (2008). "Defeating Pathogen Drug Resistance: Guidance from Evolutionary Theory." *Evolution* 62(12): 3185-3191.

Pepper, J. W., C. S. Findlay, et al. (2009). "Cancer research meets evolutionary biology." *Evolutionary Applications* 2(1): 62-70.

Pepper, J. W. (2011). Somatic Evolution of Acquired Drug Resistance in Cancer. *Targeted Therapies: Mechanisms of Resistance*. D. Gioeli. New York, Springer: 127-134.

Pepper, J. W. (2012). "Drugs that target pathogen public goods are robust against evolved drug resistance." *Evolutionary Applications* 5(7): 757-761.

Podlaha, O., M. Riester, et al. (2012). "Evolution of the cancer genome." *Trends in Genetics* 28(4): 155-163.

Ransohoff, D. F. (2009). "Promises and limitations of biomarkers." *Recent results in cancer research. Fortschritte der Krebsforschung. Progres dans les recherches sur le cancer* 181: 55-59.

Rosenfeld, S. and I. Kapetanovic (2008). "Systems biology and cancer prevention: all options on the table." *Gene regulation and systems biology* 2: 307-319.

Shah, S., R. Morin, et al. (2009). "Mutational evolution in a lobular breast tumour profiled at single nucleotide resolution." *Nature* 461: 809 - 813.

Shah, S. P., A. Roth, et al. (2012). "The clonal and mutational evolution spectrum of primary triple-negative breast cancers." *Nature advance online publication.*

Shibata, D. (2012). "Heterogeneity and Tumor History." *Science* 336(6079): 304-305.

Siegel, R., D. Naishadham, et al. (2012). "Cancer statistics, 2012." *CA: A Cancer Journal for Clinicians* 62(1): 10-29.

Sprouffske, K., J. W. Pepper, et al. (2011). "Accurate reconstruction of the temporal order of mutations in neoplastic progression." *Cancer Prevention Research* 4(7): 1135-1144.

Walter, M. J., D. Shen, et al. (2012). "Clonal Architecture of Secondary Acute Myeloid Leukemia." *New England Journal of Medicine* 366(12): 1090-1098.

In: Advances in Computational Modeling Research
Editor: Anna Belya Kora
ISBN: 978-1-62618-065-9

Chapter 6

REFLECTIONS ON AGENT-BASED MODELING: SIMULATING WEB NETWORKS

***Ioannis Katerelos*[1], *Theodore Tsekeris*[2,*] *and Charalambos Tsekeris*[1]**
[1]Department of Psychology, Panteion University of Athens, Athens, Greece
[2]Centre of Planning and Economic Research (KEPE), Athens, Greece

ABSTRACT

This article comprises a comprehensive overview of major theoretical advances in agent-based modeling as well as a computer simulation of the Web 1.0 and Web 2.0 networks. Both of these networks are represented on the basis of the graph theory and are simulated using the Holistic-Emergent Social Interaction-Oriented Dynamics (HESIOD) model. This model enables a deeper understanding and interpretation of the dynamical features, stability and self-organization properties of the networks. Several suitable measures are employed to describe the state of the systems, including the Lyapunov exponent and information entropy. The results vividly indicate that what mainly characterizes Web 2.0, and radically differentiates it from Web 1.0, is a far-from-equilibrium dynamics with lower stability and self-organization, and higher diversity. As it is demonstrated with a real-world example, these characteristics can be attributed to the explosion of the user-generated online content, that is, a fundamental bottom-up process. This process surprisingly reinforces social dynamics and substantially offers the necessary conditions for the unpredictable creativity, innovation and adaptation.

1. INTRODUCTION ON WORLD WIDE WEB AND AGENT-BASED MODELING

Transcending the static and hierarchical character of the first generation of the World Wide Web (Web 0.1), the internet of Web 2.0 is identified with ultra-complex networks exchanging or co-creating information and knowledge, ceaselessly re-shaping the social structure. Common or lay knowledge depicts Web 2.0 as the central power-engine of social

[*] E-mail address: tsek@kepe.gr (Corresponding author)

dynamics, repeatedly transmitting and organising human culture, creativity and identities. Figure 1 provides the visual representation of the Internet space world-wide, including some major IP addresses, such as net, ca, us com, org mil, gov, edu, jp, cn, tw, au de, uk, it, pl, fr br, kr, nl. The connectivity of the Internet has been found to be rapidly growing among users, attaining scale-free characteristics with high fractal dimension (Heudin, 2008) and small-world properties. These properties depict the evolutionary advantage of such architecture, since small-world networks are considered to be more robust to perturbations (for instance, due to damage by viral infection and random breakdowns) than other (e.g., hierarchical and random) network architectures (Newman et al., 2006).

In methodological terms, Web 2.0 initiated a data revolution and inspired rigorous data-driven approaches, sensitive to the dynamics of communication in virtual contexts. Nowadays, large, persistent and near real-time digital datasets pertaining to user-generated content online are searchable and almost ubiquitous. Based on the increasing interconnectedness of system components, the Web 2.0 datasets offer a huge capacity for modeling the behavior and interactions of users who know or do not know each other. These processes lead to far-from-equilibrium and highly unpredictable system effects (Barabási, 2002).

Several attempts have been made to simulate the characteristics and dynamic evolution of the Web. Goncalves and Ramasco (2008) explain that the properties of Web traffic cannot be reproduced by Markovian models, such as PageRank, in which destinations are independent of past decisions.

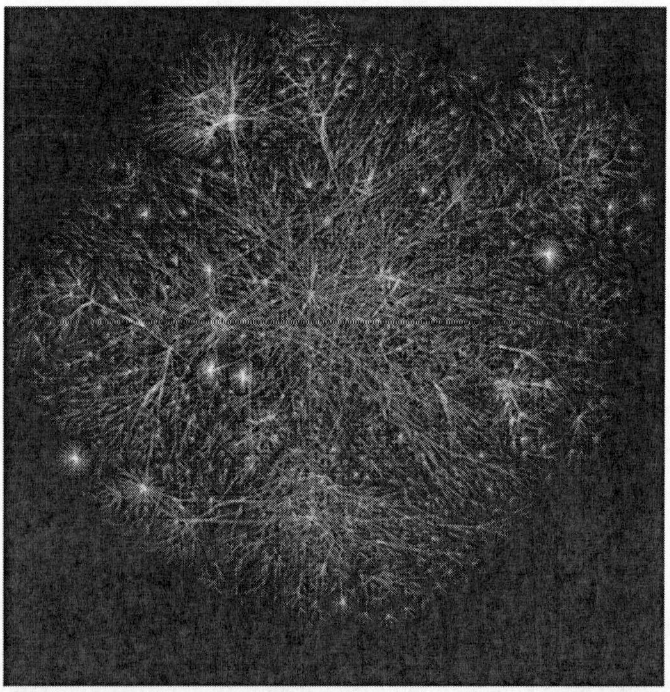

Source: opte.org.

Figure 1. Visualization of the internet space with the main IP addresses (as of 2005).

Agent-based modeling is an innovative branch of computer simulation, focusing on the modeling of human complex systems, such as those of the virtual world. It seeks to model the individual behavior, together with its unique imperfections, weaknesses, limitations, idiosyncrasies and personal interactions. Agent-based simulations can represent both the local (micro-level) and network-wide (macro-level) interrelationships among the Web users and allow the dynamical representation of bottom-up processes. Such a model with features of memory and topicality that can account for both individual and system-wide traffic patterns from extremely heterogeneous and diverse aggregate real-world traffic measurements in the Web was developed by Meiss et al. (2010). This article presents an agent-based simulation of Web 1.0 and Web 2.0 networks by using the Holistic-Emergent Social Interaction-Oriented Dynamics (HESIOD) model (Katerelos, 2012). The HESIOD model allows for the impact of both the regulation of intra-personal discrepancy and the interpersonal variability of the opinions of agents on social learning in the Web and network dynamics.

2. Comparison between the Web 1.0 and Web 2.0 Network Dynamics

Figure 2 and Figure 3 illustrate, in an abstract manner, the representation of the network (communication topology) of Web 1.0 and Web 2.0, respectively. The Web 1.0 network is preferred to be implemented as a star (one-to-all) graph, since it corresponds to a largely centrally conceived and controlled topology.

It is characterized by an individual-based type of communication, in the sense that all agents are individually linked to a central one, in a complete absence of peer-to-peer interaction. The Web 2.0 network is preferred to be implemented as a small-world graph (Strogatz, 2003), since relevant studies indicate that Web 2.0 exhibits small-world features (Barabási, 2002). Such network types present a relatively high connectivity, in terms of the average clustering coefficient, and relatively low average shortest path, although they have a relatively low density and are rather scarce, i.e., with only few direct connections between agents.

The current set of simulations is experimentally designed and implemented with a population of 100 agents for each network, in order to obtain a deeper understanding of the properties of the system dynamics. The HESIOD model, which seeks to systematically quantify social dynamics, employs two basic indexes: Lyapunov exponent (λ) (Sprott, 2003) and information entropy (I) (Peitgen et al., 1996). The former measures the system's sensitivity to initial conditions, thus expressing the overall stability or instability of the network. The latter expresses the degree of self-organisation of the system, since less entropy entails more order, less diversity and less chaos.

Figure 4 and Fig 5 depict the Web 1.0 and Web 2.0 network dynamics, respectively, using the HESIOD model for 20,000 iterations (note that agents are "equipped" with a balanced personality: K=1 and Ψ=1, see Katerelos 2012). Moreover, Figure 6 and Figure 7 describe the dynamic evolution of the Web 1.0 and Web 2.0 entropy measures and Lyapunov measures, respectively. For the Web 1.0 network, Figure 4 shows that, after a short period of turbulence (transient chaos), the system stagnates to a fixed point (flat lines). The Lyapunov exponent here equals to λ=0.001 (Figure 7), which clearly signifies that the Web 1.0 network is mostly a top-down system. Namely, it is static and does not essentially fuel social dynamics. On the contrary, such a network tends to stabilize social dynamics. Also, the Web

1.0 network exhibits a small entropy I=2.53, which implies that such systems lead the artificial society to a stable and highly ordered (self-organised) state.

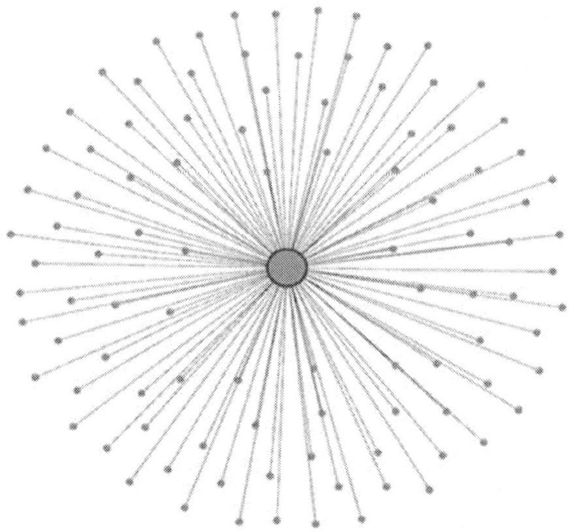

Figure 2. Web 1.0 network topology representation as a star graph.

Figure 3. Web 2.0 network topology representation as a small-world network.

For the Web 2.0 network, the experimental results are quite different. The system is in permanent movement, without any steady state, as reflect the network dynamics (Figure 5) and the highly fluctuated temporal evolution of the entropy (Figure 6) and Lyapunov (Figure 7) measures. Specifically, at the end of the analysis period (20,000 iterations), the Lyapunov exponent is now λ=0.150, which denotes high sensitivity to initial conditions, while the information entropy is I=3.24, which means that the system is moderately ordered. Namely, the Web 2.0 network, which can be considered as a bottom-up system, demonstrates lower stability, less self-organization and higher diversity, compared to the Web 1.0 network. These findings verify that Web 2.0, as a complex and unpredictable system, indeed reinforces social dynamics and can rapidly produce social facts, thus introducing a stable flow of unpredictable creativity, innovation and adaptation. Through the development of the Web 2.0, social groups become continuously re-shaped and de-shaped, signifying that the system is constantly organizing and dis-organising itself, never reaching a state of equilibrium.

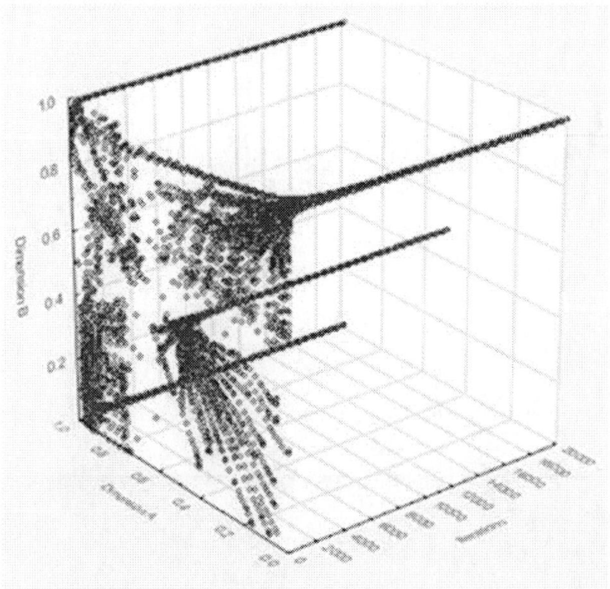

Figure 4. Web 1.0 network dynamics using the HESIOD model (20,000 iterations).

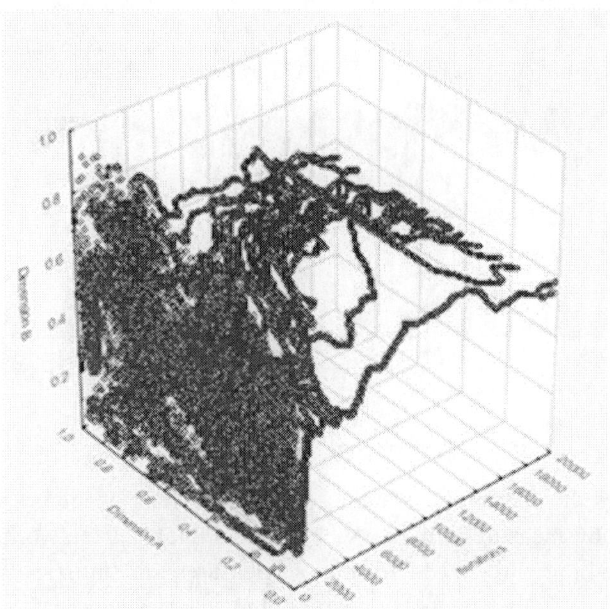

Figure 5. Web 2.0 network dynamics using the HESIOD model (20,000 iterations).

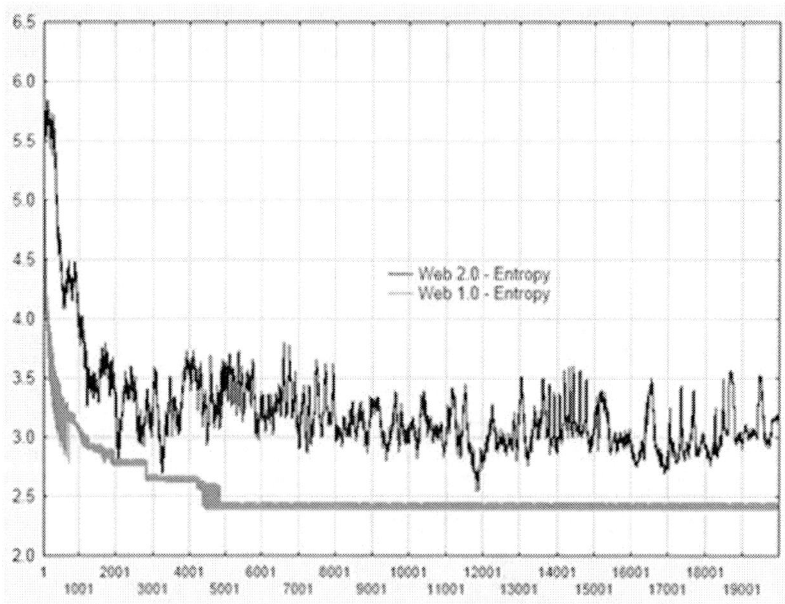

Figure 6. Dynamic evolution of Web 1.0 and Web 2.0 information entropy measures.

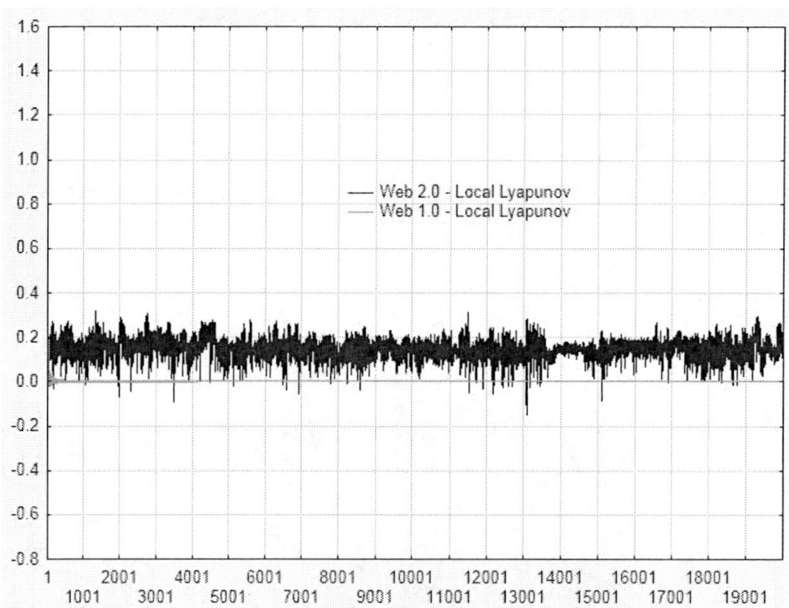

Figure 7. Dynamic evolution of Web 1.0 and Web 2.0 Lyapunov measures.

Table 1 provides the descriptive statistics and the correlation matrix of the information entropy and Lyapunov measures for the Web 1.0 and Web 2.0 networks. In the Web 1.0 network, the information entropy and Lyapunov measures are statistically significantly and moderately correlated. This outcome denotes the co-existence of stability and self-organization of the system over time. On the contrary, in the Web 2.0, the negligible magnitude of correlation between the information entropy and Lyapunov measures indicates that the instability and shortage of self-organization do not persistently co-exist over time. Finally, the information entropy measures of Web 1.0 and Web 2.0 are highly and significantly positively correlated, in contrast with the very small and negative correlation of Lyapunov measures. This outcome signifies the existence of a similar deterministic time trend in the self-organization of the two systems. However, in the Web 2.0, the entropy presents high fluctuations over time, while, in the Web 1.0, the entropy is very quickly stabilized. The following section describes the role of social media on expanding the information diversity and stimulating emergent dynamic social processes in the Web 2.0.

Table 1. Descriptive statistics and correlation matrix of entropy and Lyapunov measures for the Web 1.0 and Web 2.0 networks

	Means	Std.Dev.	Web 2.0 Lyapunov	Web 2.0 Entropy	Web 1.0 Lyapunov	Web 1.0 Entropy
Web 2.0 Lyapunov	0.1497	0.0423	1.0000	**0.0241**	**-0.0169**	**0.0933**
Web 2.0 Entropy	3.2409	0.4445	**0.0241**	1.0000	**0.1591**	**0.8239**
Web 1.0 Lyapunov	0.0007	0.0179	**-0.0169**	**0.1591**	1.0000	**0.2493**
Web 1.0 Entropy	2.5283	0.2709	**0.0933**	**0.8239**	**0.2493**	1.0000

Note: Bold marked correlations are statistically significant at $p < .0500$ (N=20,000).

3. Modeling Emergent Social Processes through the Web 2.0

Social media can be defined as a group of Internet-based applications that rely on the ideological and technological foundations of Web 2.0 and allow the creation and exchange of user-generated content (de Belt et al., 2012). They offer to users several unforeseen opportunities to participate in online social networks, using highly accessible and scalable communication applications. One of these applications refers to both individualistic and collective protest activities based on user-generated content. Synergetic creation processes such as posting, classifying or evaluating content online explicitly pose new challenges regarding the actual shaping of cultural and social practices, as well as of political connections and dynamics. These processes are crucial to systematically understand and model the multiple ways in which Web 2.0 is smartly used for protest activities and fast changing the political landscape (Breindl, 2010).

During the last years, social media have been famously connected to different forms of civil disobedience and collective action, such as the Arab spring, the Occupy Wall Street movement, student protests, and social movements in Greece, Portugal, Spain, and other countries. Subsequently, there is a growing amount of computational research that vividly shows how these platforms have profoundly affected democracy, civic engagement and collective action, social sharing and coordination, human communication and relationships, privacy, etc. In addition, state power and politics are intriguingly both contested and exercised upon social media. In a theoretical context, social media are increasingly serving both as an effective tool and as a multi-layered environment or terrain for disputes and conflicts between groups, states, or a plethora of organised and autonomous collective agents. They substantially empower otherwise powerless, voiceless or oppressed actors, thus exercising a unique form of technopolitics.

In the Arab spring, thousands of activists promptly and imaginatively used social media as an alternative organizing, information and communication tool in their protests. Figure 8 illustrates the social dynamics of the Egyptian revolution (at the time of the announcement of Mubarak's resignation on February 11, 2011) as a virtual event. The network nodes are Twitter users, who rapidly responded in real time through the Web; the few nodes and edges, sparse in time, were dramatically increased with the receipt of the announcement. Figure 8 is the outcome of the dynamic network representation of the boom of retweets with the use of Gephi open-source software for visualizing and analyzing large networks. In addition, Twitter vastly triggered, facilitated and empowered the so-called "Green Movement" during the 2009 elections in Iran. Excluded or marginalised minorities, ideological parties, cultural and political groups are now able to dynamically enrich, promote and negotiate their particular agenda on free-access and easy-to-use virtual platforms. The current collective discourses, such as those of "YouTube democracy" and "Twitter public sphere", enthusiastically inspire and promote social media politics, but also considerably reinforce popular techno-optimistic approaches claiming that social media help to synergetically invigorate, reconstitute or revive democracy.

Source: gephi.org/2011/the-egyptian-revolution-on-twitter.

Figure 8. Network representation of the Egyptian revolution on Twitter.

Nevertheless, social media sometimes lead to asymmetrical power relations, directly yielding from asymmetrical information relations of online visibility. Some governments have systematically attempted to censor and control social media with the help of advanced surveillance technologies. Furthermore, the omnipresence of cybercrime, cyberharassments, cyberterrorism or cyberextremism, and various discourses on the multiple ways in which the police, intelligence, or repressive regimes monitor social media, repeatedly fuel techno-pessimistic approaches that view social media as a new serious threat to contemporary democracy.

Conclusion

Since the mid-1990s, we can discern various examples and computational methods to show how the Internet has been expanded and used to substantially promote social network dynamics. As it was shown from the present agent-based simulation experiments, Web 2.0 involves far-from-equilibrium dynamic processes with lower stability, lower self-organization and higher diversity. Hence, in the highly changing and risky context of complex virtual topologies in the Web, any 'top-down' control of information spread and opinion formation is completely impossible and undesirable. Equally undesirable is a linear, predictable, stable, ordered and pure homogenous world.

Agent-based modeling substantially helps us to carefully analyze the behavior of the consumers of real/virtual worlds and their local interactions. What we actually see is the tremendous impact of these interactions on global patterns, and conversely, the implications of global structures for local behaviors. We can subsequently synthesize these worlds, run simulation experiments and explore intuitions and 'what-if' scenarios. Understanding then comes from emergence, chaos and diversity.

Recent developments in agent-based modeling are just a promising starting point, not an end. The future dynamic evolution of the Web can be coarsely projected up to a certain time horizon (i.e. *predictability horizon*), but it cannot be fully predicted with certainty and precision in the long run.

References

Barabási, A.-L. (2002) *Linked. The New Science of Networks*, Perseus, Cambridge, MA.

Breindl, Y. (2010) Critique of the democratic potentialities of the Internet: A review of current theory and practice. *tripleC*, 8(1), pp. 43-59.

Giglietto, F. and Rossi, L. (2012) *Ethics and interdisciplinarity in computational social science*, *Methodological Innovations Online*, 7(1), pp. 25-36.

Gonçalves, B. and Ramasco, J.J. (2008) Human dynamics revealed through Web analytics. *Physical Review E*, Vol. 78, No. 2, p. 026123.

Heudin, J.-C. (2008) *Les Créatures Artificielles: des Automates aux Mondes Virtuels*, Odile Jacob, Paris.

Katerelos, I. (2012) *Chaos and Order in Social Systems*. The HESIOD Model: Social Dynamics Simulations. Papazisis, Athens.

Meiss, M., Gonçalves, B., Ramasco, J., Flammini, A. and Menczer, F. (2010) Modeling traffic on the Web Graph. *Lecture Notes in Computer Science*, Vol. 6516, *Algorithms and models for the Web-Graph*, Springer, Berlin, pp. 50-61.

Newman, M., Barabasi, A.-L., and Watts, D.J. (2006) *The Structure and Dynamics of Networks*, Princeton University Press, Princeton, NJ.

Peitgen, H-O., Jurgens, H. and Saupe, D. (1992) *Chaos and Fractals: New Frontiers of Science*, Springer, NY.

Sprott, J.C. (2003) *Chaos and Time-Series Analysis*, Oxford University Press, Oxford, UK.

Strogatz, S. (2003) *Sync: How Order Emerges from Chaos in the Universe, Nature and Daily Life*, Hyperion, NY.

Van de Belt, T.H., Berben, S.A.A., Samsom, M., Engelen, L.J. and Schoonhoven, L. (2012) Use of social media by Western European hospitals: Longitudinal study. *Journal of Medical Internet Research*, 14 (3), e61 (http://www.jmir.org/2012/3/e61/).

In: Advances in Computational Modeling Research
Editor: Anna Belya Kora
ISBN: 978-1-62618-065-9

Chapter 7

Model-Based Approaches to Diagnosis of Multi-agent Plans: A Survey

Roberto Micalizio*
Dipartimento di Informatica, Università di Torino,
corso Svizzera Torino, Italy

Abstract

Diagnosis is an essential step in the process of making robust complex systems and in particular Multi-Agent Systems (MASs). This Chapter reviews the most important model-based techniques to diagnosis of MASs that have been proposed over the last decade.

Since the term MAS refers to a wide range of systems, the Chapter focuses on MASs that can be represented as Multi-Agent Plans (MAPs), and categorizes the approaches to diagnosis of MAPs along different dimensions. Two features are particularly important to characterize the examined methodologies: the MAP model, which can either be implicit or explicit, and the notions of diagnosis (i.e., social, plan, or agent diagnosis). This categorization enables a direct comparison between the different solutions highlighting their advantages and disadvantages, as well as their similarities.

The Chapter ends with some remarks about future developments on Model-Based Diagnosis (MBD) applied to MAPs.

Keywords: Model-Based Diagnosis, Multi-Agent Plans, Teamwork

AMS Subject Classification: 68T42, 68T20, 68T30

1. Introduction

Multi-Agent Systems (MASs) are nowadays employed in many humans' activities. With the term MAS we refer to a wide variety of systems, either software or physical, characterized by a number of agents acting simultaneously in the same environment. Agents can either be *self-interested* – as in e-commerce where a seller tries to maximize its profits

*E-mail address: micalizio@di.unito.it

and a buyer tries to minimize its bills – or *cooperative* – as for instance in the RoboCup [1] where the robots in a team work together to get a common result: win the match.

Whether they are software or physical, a common problem in dealing with MASs is how to make them sufficiently *robust* to be employed in real-world domains. Differently from synthetic domains, in fact, the real world changes unpredictably, and these changes might have dramatic effects on the activities carried on by the agents. This characteristic poses critical problems especially in cooperative teams, where the failure of one agent can cause the failure of many others. To prevent, or at least mitigate, the impacts of unexpected changes, the team members *must adapt* themselves to the new situations happening in their environment.

One of the critical steps involved in such an adaptation process is *diagnosis*. Diagnosis, in fact, is the task of detecting and explaining anomalous results and situations. The rationale is that, only when the agents can identify what went wrong and why, they can intervene properly by adapting themselves to unexpected events.

This Chapter reviews the main model-based approaches to diagnosis of MASs that have been discussed in the last decade. Since the corpus about this subject is rather wide, we narrow the scope of this Chapter to systems of cooperating agents whose activities can be organized into a Multi-Agent Plan (MAP). Our purpose is to highlight the main features characterizing the different solutions proposed in literature, and categorize them along different dimensions. For instance, the models used to represent actions and agents vary significantly from one approach to another. Additionally, in some cases diagnosis is solved by a single diagnostic agent, whereas other approaches advocate a distributed methodology. This variety of approaches is due to the fact that they are designed to solve the diagnostic task in specific scenarios. Knowing the range of possible solutions, one can decide what is the best one to be applied (or even extended) in the particular domain at hand.

The Chapter is organized as follows. In Section 2 we briefly review the main concepts about Model-Based Diagnosis (MBD). In Section 3 we compare two ways of thinking about MASs either as societies of agents, or as distributed systems. This distinction prepares the ground for the first main categorization we make in Section 4, where we individuate two main families of approaches depending on whether the MAP is modeled explicitly or not. In *teamwork-oriented* approaches, the MAP is not explicitly modeled, but the relations among the agents are weakly modeled via a teamwork model. On the other hand, in *action-oriented* approaches, the MAP is explicitly modeled in terms of action instances. Approaches in these two families are discussed in sections 5 and 6, respectively. In Section 7 we summarize the main features we have identified along the Chapter. Finally, in Section 8, we draw some possible future directions in the field, and then conclude.

2. Background on Model-Based Diagnosis

Diagnosis has drawn the attention of the Artificial Intelligence (AI) community since early 1980s. The reason of this interest must be sought in the attempt of diagnosing automatically, and efficiently, *complex systems*; that is, systems consisting of a huge number of components related to one another in different ways.

Intuitively, diagnosis is the task of explaining why a given system has got wrong results. A characteristic that makes the diagnostic task challenging is that we are usually interested in finding *all* the possible explanations. The search space, however, is exponential in the number of system's components [48]. Moreover, once we have computed all the possible diagnoses, these diagnoses must be ranked according to some preference criteria in order to put first explanations that are more preferable (or more likely) than others.

In this section, we briefly summarizes the main concepts about Model-Based Diagnosis (MBD) [24], as the notions of consistency-based and abductive diagnosis, that are at the basis of many model-based approaches to the diagnosis of MAPs.

Consistency-Based Diagnosis. The model-based approach to the diagnosis has its roots in the seminal work by Davis [13]. Davis highlights the importance of representing the knowledge about the system structure and about its behavior. The basic idea of MBD is to exploit a model of the *nominal* behavior of the system in order to build expectations about its outputs knowing its inputs. A diagnostic problem arises when the outputs actually observed differ from the expected ones. The diagnostic process has to single out which components in the system behaved abnormally and caused the erroneous observed outputs.

A significant contribution to MBD was given by Reiter. In his work about diagnosis from first principles [48], Reiter presents a first formalization of the diagnostic problem using first-order logic for encoding the system model. In the rest of this section we recall the formalization proposed by Reiter as it is the basis for many of the other formalizations proposed within the MBD community.

Definition 1 (System Model). *A system is a pair* $(SD, COMPS)$ *where:*

- SD *is the system description consisting of a set of first-order sentences;*
- $COMPS$ *is the finite set of system components.*

In the early view by Reiter, the system model describes only the nominal behavior of the system itself. To this end, Reiter introduces the special predicate $\text{AB}(\cdot)$: When the predicate $\text{AB}(c)$ holds for a given component $c \in COMPS$, then the component c is assumed to function abnormally, hence its outputs cannot be nominal. Having the $\text{AB}(\cdot)$ predicate, we can easily model the nominal behavior of a component as first-order sentences. For instance, the correct behavior of an AND gate in a digital circuit is given by the formula:

$$\text{AND}(x) \wedge \neg\text{AB}(x) \supset \text{out}(x) = \text{and}(\text{in1}(x), \text{in2}(x)).$$

Note that this formula is instantiated for each and-gate component c in $COMPS$.

A diagnostic problem arises when, given a set of observations OBS about the system behavior, we detect a discrepancy between OBS and the expected behavior predicted by the system model. Observations are again sentences in first-order logic; formally:

Definition 2 (Diagnostic Problem). *A diagnostic problem is a triple* $(SD, COMPS, OBS)$ *such that*

$$SD \cup \{\neg\text{AB}(c_1), \neg\text{AB}(c_2), \ldots, \neg\text{AB}(c_{|COMPS|})\} \cup OBS \models \bot.$$

This means that, given the observations in OBS and the system model $(SD, COMPS)$, it is not consistent to assume that all the components in $COMPS$ behaved correctly: For at least one component in $COMPS$ the predicate $\text{AB}(\cdot)$ must be true.

Resolving a diagnostic problem requires to restore the consistency by assuming the abnormality for a (minimal) subset of system components.

Definition 3 (Consistency-Based Diagnosis). *A consistency-based diagnosis for the problem* $(SD, COMPS, OBS)$ *is a minimal set* $\Delta \subseteq COMPS$ *such that*

$$SD \cup OBS \cup \{\textsc{ab}(c)|c \in \Delta\} \cup \{\neg\textsc{ab}(c)|c \in COMPS - \Delta\} \text{ is consistent.}$$

In other words, Δ is a diagnosis iff it is sufficient to assume that just the components in Δ behaved abnormally (while all the others behaved correctly) to restore the consistency between SD and OBS.

It is important to note that, when the observations match the expectations, then the only possible diagnosis is $\Delta = \emptyset$; in fact, only when all the components behave nominally we obtain the expected output from the system.

The consistency-based approach proposed by Reiter is complemented by the work by de Kleer et al. [15]. In particular, de Kleer et al. view the diagnosis as an incremental process guided by measurements that, taken during the diagnostic inference, refine (i.e., reduce) the space of diagnostic hypotheses.

Abductive Diagnosis. A drawback of the consistency-based diagnosis is that it just selects a subset of components that cannot be nominal, but it does not tell us nothing about the reason for the abnormality of these components. In some cases, however, one is in interested in knowing the specific abnormal behavior assumed by a component (see e.g., [37, 38] where agent diagnosis drives the recovery from an action failure). This information cannot be inferred in a consistency-based approach because of the abnormal predicate AB, which abstracts all the possible anomalous modes of a components.

To infer abnormal behavioral modes, the model of the system must be enriched. The model must include not only the nominal behavior, but also the anomalous behavior of each system component. More precisely, for each component in the system, one has to specify *all* the possible ways (i.e., modes) in which that component may fail. Thus, for each abnormal mode of each component, the system description must specify what are the outputs produced by that component behaving in that specific abnormal mode.

For instance, let us consider again an AND gate, and let us assume that this type of gate can be affected by an anomalous mode *stuck-zero*. When an AND gate behaves this mode, it produces a zero in output independently of its inputs. This mode could be modeled as follows:

$$\textsc{and}(x) \wedge \textit{stuck-zero}(x) \supset \text{out}(x)= 0.$$

Relying on this explicit representation of abnormal modes, Console et al. [11] have proposed the notion of *abductive diagnosis*. First of all, a diagnostic candidate Δ is an assignment of (possibly faulty) behavioral modes to all the system components. Thus each component in $COMPS$ is qualified by the candidate hypothesis Δ either as *nominal* or as in a specific abnormal mode.

Definition 4 (Abductive Diagnosis). Δ *is an abductive diagnosis for the diagnostic problem* $(SD, COMPS, OBS)$ *iff it contains a minimal subset of components in* $COMPS$ *that are qualified as non-nominal such that:*

1. $SD \cup OBS \cup \Delta \not\models \perp$,

2. $SD \cup \Delta \cup \models OBS$.

The first condition is analogous to the consistency-based definition of diagnosis. The second condition imposes that the diagnosis is not only consistent with the observations and the system description, but it also allows us to infer, or predict, the observations.

Indeed, in [11], the authors argue that the diagnosis Δ could also be hybrid. In their spectrum of logical definitions, in fact, they state that in general one is interested in explaining abductively just a subset of the observations in OBS; whereas, the remaining observations can be explained just in the consistency-based way. The set of observations is therefore partitioned into two subsets OBS^{abd} and OBS^{con}. The diagnostic problem becomes a quadruple $(SD, COMPS, OBS^{abd}, OBS^{con})$, meaning that the observations in OBS^{abd} must be explained abductively, while for the observations in OBS^{con} it is sufficient to prove their consistency. In such a general framework, a solution Δ must satisfy the following definition.

Definition 5. *A candidate diagnosis* Δ *is a solution for the diagnostic problem* $(SD, COMPS, OBS^{abd}, OBS^{con})$ *iff:*

1. $SD \cup OBS^{con} \cup \Delta \not\models \perp$,

2. $SD \cup \Delta \cup \models OBS^{abd}$.

Of course, when OBS^{abd} is empty, Δ must just be consistent with all the observations, and hence we fall back into the pure consistency-based diagnosis – one extreme of the spectrum. On the other hand, when OBS^{con} is empty, Δ must entail all the observations, so we fall back into the pure abductive diagnosis – the other extreme of the spectrum.

Temporal Diagnosis. The spectrum of logical definitions proposed by Console et al. has also been extended to consider the temporal dimension. In [5], Brusoni et al. present an ontology of temporal phenomena that are relevant for the diagnostic task, and according to such an ontology they extend the notions of abductive and consistency-based diagnosis. To keep the discussion simple, however, we do not address the logic formalization of temporal diagnosis.

The Model-Based Principle. It is important to note that, although the formalization introduced by Reiter (and subsequently developed in [11, 5]) was based on first-order logic, the MBD approach is independent of the language used for expressing the system model. As we show in sections 5 and 6, in fact, the diagnosis of MAPs has been attempted with the adoption of models expressed in different types of formalisms and languages. The main advantage of a model-based approach is that both the system model and the diagnostic inferences can be tailored on the type of diagnoses one is interested in. In other words, one first thinks about the information she/he wants to obtain from a system, and then build a system model accordingly.

3. Agent Societies *vs.* Distributed Systems

One of the main concerns in dealing with Multi-Agent Systems is how to make them robust when operating in the real-world. In real-world scenarios, in fact, unexpected contingencies can arise at any time. Contingencies can be due to events occurring in the environment (e.g., a node of the Internet may crash), or to harmful interactions among the agents themselves that unpredictably arise in complex and dynamic situations (e.g., incorrect use of critical resources).

In this section we review some of methodologies that, around the end of 1990s and the early 2000s, have coped with the problem of making a MAS robust, and hence have paved the way for the subsequent development of MBD in teams of cooperating agents. These approaches can be roughly subdivided into two main categories: the approaches based on an explicit representation of the *teamwork*, and the approaches based on *distributed systems*.

Solutions based on teamwork highlight the need to represent how the members of a team can correctly coordinate among themselves. The team is seen as a society where agents must respect some norms of behavior. When these norms are not respected, the team is in a disagreement meaning that some problem has occurred. To identify the reason of such disagreements, the research on teamwork has borne the *social diagnosis* methodology that we discuss in detail in Section 5.

Solutions based on distributed systems are somewhat dual to the previous ones. Distributed systems are typically used to cope with complex problems by decomposing these problems into subproblems. Approaches in this category are focused on what each agent in the team does to solve a subproblem. The coordination with other teammates is more constrained than in the agent society case as it is usually established at design time (i.e., in the process of decomposing the original problem into subproblems). The attention here is on the achievement of goals and subgoals. Distributed systems have borne a family of *action-oriented* methodologies where the diagnostic process infers a *plan diagnosis*; that is, a subset of plan actions qualified as abnormal. Action-oriented approaches are addressed in Section 6.

3.1. Diagnosis of Teamwork Failures and Agent Societies

Teamwork Failures. First attempts in dealing with *teamwork failures* have been proposed by Tambe et al. [57, 35]. Tambe highlights that teamwork is something more than the coordinated execution of actions carried on simultaneously by a team of agents. Teamwork also requires the team members a deliberative effort in achieving a common goal; in short, teamwork is coordination plus cooperation.

Having this in mind, Tambe suggests that in order to build a robust multi-agent system, agents must be provided not only with their own goals and plans, but also, and more importantly, with an explicit representation of their teamwork. Such a model must outline the commitments of each member in the team, and its responsibilities as a participant of the team in a specific role. Relying on a teamwork model, an agent can appropriately select its own actions in accordance with the actions selected by others and without creating hazardous situations for the rest of the team and for itself.

Tambe builds up on the *joint intention framework* introduced by Cohen and Levesque

[10]. In this framework a team jointly intends a *team action* if all the team members are jointly committed to completing that action. Thus, all the team members are simultaneously pursuing a *joint persistent goal* (JPG). A JPG about a goal p holds if three conditions are satisfied:

1. all members mutually believe that p is currently false;
2. all members mutually know that they want to get p;
3. all members mutually believe that until p is mutually known to be achieved, unachievable or irrelevant, they mutually believe that each of them holds p as a *weak goal*.

A *weak goal* is just the individual goal of a team member. It either represents the intention of an agent to get that weak goal (e.g., p is currently believed false, and the agent wants p eventually to be true), or the intention of an agent to drop p since it has discovered that p is no longer relevant or achievable. In this second case, the agent has to propagate its local point of view to all the other members of the team.

Tambe suggests to implement the joint intentions framework by means of a hierarchy of *team operators* that can be decomposed into individual operators and plans. Team operators, as well as individual operators, are qualified by:

- precondition rules for their selection;
- application rules;
- termination rules.

Note that team operators apply to *team states*; these are abstractions of the mutual beliefs of the team members about the world, including information about their joint tasks and intentions. Thus, precondition, application, and termination rules are all conditions on team states.

A big issue, of course, is how such team states should be inferred since they involve all the members of the team. A solution in which each agent communicates with all of its teammates is unacceptable. First, because the number of messages exchanged among the agents grows exponentially with the size of the team. Second, in some situations agents might not be allowed to communicate at all. To mitigate this problem, Tambe suggests that one (or more) member(s) in the team is in charge of tracking the other team members in order to guess their beliefs (see for instance the methodology in [34]).

Figure 1 sketches an example[1] of the team operators hierarchy proposed by Tambe. This example is taken from the ModSAF domain [8]. In the ModSAF domain agents are helicopters involved in a military-like mission. Team members are subdivided into two roles: the attackers and the scouts. A mission requires the scouts to go ahead and individuate the target, while the attackers awaits in safe positions. Once the target has been located, the attackers have to be activated by the scouts.

Figure 1 shows how a mission is decomposed into a team operators hierarchy: box-nodes represent team operators, each team operator can be decomposed into other team

[1] The picture is a simplified version of Figure 4 in [57].

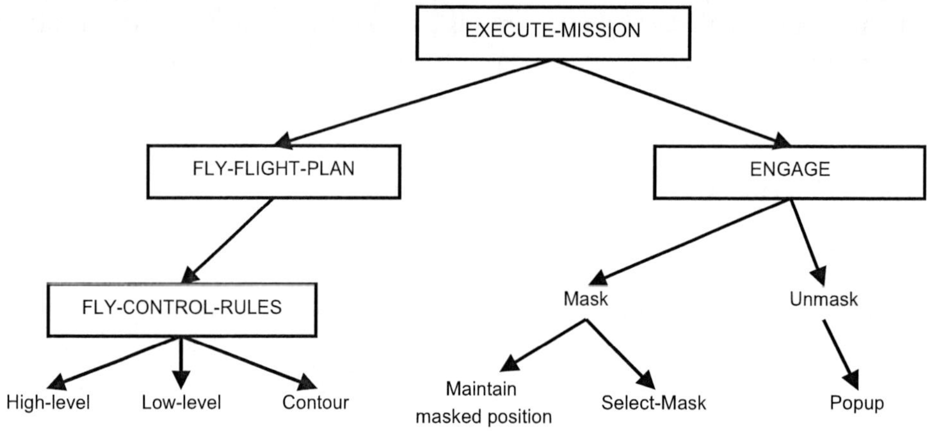

Figure 1. A portion of the team-operator hierarchy for the ModSAF domain in [57].

operators or individual plans, which in turn are decomposed into individual activities (the leaves of the hierarchy).

Thus, the basic idea is that having such a hierarchy of team operators, the team members can synchronize in order to select the same operator. The synchronized selection of a team operator causes the team members to undertake either another team operator or an individual plan according to how the selected team operator decomposes within the hierarchy.

Tambe extends the joint framework by introducing the notion of roles. A role is a means to constraint the behavior of an individual team member or of a subteam. A role in fact drives the decisions that an agent takes about its next activities. It is important to note that the performance of an individual agent in playing its role is not necessarily determinant for the successful completion of the mission of the whole team. The accomplishment of the mission depends, in fact, on the ability of the agents to adapt themselves in playing different roles. For instance, if an agent playing a critical role crashes, the mission will be aborted unless there exists another agent that can substitute the crashed agent in playing that critical role.

Although Tambe does not explicitly address the problem of diagnosing the teamwork, he poses the basis for the subsequent development of *social diagnosis* by Kalech and Kaminka that we discuss in the next section.

Agent Societies. While Tambe [57] focuses on the explicit representation of teamwork, Dellarocas et al. [20] see a MAS as an open society where agents can freely join or leave. They consider software agents operating in the Internet. Each of these agents has to cope with unreliable infrastructures (e.g., nodes can crash), non-compliant agents (i.e., agents are heterogeneous and there are no guarantee that they are "honest"), and erroneous, unexpected interactions with other agents.

In such a scenario it is no longer possible to talk about teamwork. Agents in fact are in general self-interested: they do not cooperate with one another, they rather coordinate with others in order to reach their own objectives. Coordination is gained via protocols that specify roles and rules that the involved agents have to play and satisfy. Unfortunately, it is very hard to specify coordination protocols that anticipate all the possible exceptions that might occur at runtime.

Dellarocas et al., in fact, point out that extending coordination protocols with extensive exception handling strategies has a number of severe drawbacks. Besides the difficulty in anticipating all the possible exceptions, the design of agents becomes more complex and hence they are more difficult to maintain over time. Moreover, when protocols become too complex, it may difficult to guarantee that an agent is actually compliant with them. Consequently, not all the exceptions could be properly handled by the agents. Finally, the performance of the system may drop significantly as the agents waste time in avoiding exceptions that would not occur.

To overcome the problem of overweighted coordination protocols, Dellarocas et al. propose the introduction of specialized agents in charge of handling exceptions. In their view a MAS is composed of two classes of agents. One is the class of agents that solve users' problems, these agents are called "citizens". The other class of agents has to guarantee the correctness of the operations performed by the citizens. In analogy with humans' societies, this second class of agents represents social institutions (such as the police) that by virtue of their widely accepted legitimacy can intervene in the citizens' activities. This solution simplifies the design of the citizens as they can just assume an optimistic behaviors (e.g., an agent can trust in other agents). While the specialized, institutional agents have to surveil the citizens in order to detect potentially fraudulent agents and stop them.

The solution proposed by Dellarocas et al. has the advantage to model a MAS as an open society where agents of different nature and with different purposes can join and leave freely. The adoption of institutional agents, however, has some limitations. The problem of dealing with exceptions has in fact been moved from the citizens to the institutional agents. For instance, in a huge and dynamic society of citizens, more institutional agents have to coordinate with one another in order to guarantee that the whole system is properly surveilled, but institutional agents may fail themselves; thus they could be involved in problems similar to the ones they try to solve for the citizens.

3.2. Diagnosing Distributed Systems

Model-based approaches to the diagnosis of complex, spatially distributed systems have been proposed since the early 2000s (see e.g., [22, 17, 23, 51]). The starting point of these methodologies is that, in many real-world applications, it is not possible, or convenient, to synthesize a single model of the entire system under consideration. For instance, telecommunication and computer networks typically involve a huge number of components, and a single model would inevitably be awkward to deal with. Moreover, a centralized approach would probably be ineffective as events observed in a remote site should be transmitted to a single diagnostic agent, and the communication latency might not be negligible (see e.g., [47]).

To deal with huge complex systems, the approaches in [22, 17, 23, 51] propose to distribute the knowledge about the system among a number of cooperating *diagnostic agents*. Each diagnostic agent is therefore capable of diagnosing a portion of the global system relying on its local knowledge, and when required (e.g., when it cannot isolate a fault within its subsystem), it can cooperate with others in order to get further information. As pointed out by Fröhlich et al. in [22], there are two main ways for distributing the knowledge about a system. Knowledge can either be "decomposed" *semantically* or *spatially*.

In semantically distributed knowledge, each diagnostic agent has a model of whole system, but the model just considers a single aspect. For example, an agent has the model of the plumbing of a building; whereas another agent has the model of the power units of the same building. Semantic distribution is hard to handle when two or more agents need to cooperate. Each diagnostic agent has in fact a specialized knowledge that is not necessarily meaningful for the other diagnostic agents. Thus it is not clear which pieces of information the agents could exchange with one another in order to get a global agreement about the global diagnosis of the system. Due to this inherent complexity, most of the approaches we discuss in the following distribute the system knowledge spatially only.

In spatially distributed knowledge, the system is ideally decomposed into subsystems. Each subsystem is modeled by a specific model assigned to a diagnostic agent. Note that in this case the model covers all the aspects of the subsystem not just one as in the semantic decomposition. Each agent is able to diagnose locally its own subsystem, and when it cannot locate a faulty component its starts a cooperation with other agents. To ease the cooperation, the model of a subsystem specifies not only what are the components within that subsystem and how these components are connected one other, but it also specifies how the subsystem connects to other subsystems. This allows an agent to knows which other agents can be helpful in solving the diagnosis of its own subsystem, and to determine which pieces of information should be exchanged among the cooperating agents. Connections in fact are usually represented as variables shared by different agents, so that they can exchange information about these variables in order to get a better picture of their own subsystem.

The early work by Fröhlich et al., however, relies on some critical assumptions. First of all, they assume that the local diagnoses inferred by an agent about a subsystem are always reliable. Moreover, the connection between two subsystem is directed so that one subsystem can influence the other, but not vice versa. Finally, the connections among the subsystems do not create loops. Relying on these assumptions, a diagnosis of the whole system can be achieved by composing the local results inferred for each subsystem. In general, however, subsystem diagnoses that are locally consistent are not guaranteed to be globally consistent. Subsequent works in literature, in fact, have shown that when we release some of these assumptions inferring a global diagnosis becomes harder.

Undirected and circular connections have been considered in [49]. Since in this case a subsystem can influence and being influenced by its neighbors, global diagnoses could consist of a loop of blames. In fact, since each agent cannot individuate a faulty component in its own subsystem, it assumes that the faulty component must be located in some of its neighboring subsystems. Although local diagnoses are consistent, the resulting global diagnosis is not meaningful. The loop of blames can only be avoided by imposing a minimal amount of observations especially about the connection points between two subsystems.

The problem of inferring a global diagnosis from the cooperation of a team of diagnostic agents is also considered in [50, 51]. These works present coordination protocols that drive the diagnostic agents during their inferences. These protocols guarantee that the local diagnoses inferred by each agent are globally consistent, even though there is no guarantee that the global diagnosis obtained by combining local ones is minimal.

In some cases, the diagnostic agents could be provided with incomplete or erroneous knowledge about the subsystems they are responsible for. This aspect is addressed in [52] where a protocol to get an agreement on the global diagnosis is presented.

Other works (see e.g., [47, 17, 16, 2]) about distributed diagnosis are specifically designed to deal with Discrete Event Systems (DESs), and fall outside the scope of this Chapter.

4. Diagnosing Multi-agent Plans

As stated in the introduction, this Chapter reviews the main model-based approaches to the diagnosis of multi-agent plans proposed in the last decade. These approaches differ one another in different ways; however, relying on the distinction we have made in the previous section between *agent societies* and *distributed systems*, we can draw a first, coarse categorization and individuate two main families of approaches: *teamwork-oriented* and *action-oriented.*

In *teamwork-oriented* approaches, the plan assigned to the team members is represented *implicitly*; that is, it is not important what actions the agents are going to perform, but it is important how the agents are going to interact with one another. In *action-oriented* approaches, on the other hand, the plan actions are represented *explicitly*, and the coordination/cooperation among the agents is represented in terms of services exchanged between actions of different agents.

This initial distinction helps us to intuitively define a diagnosis for a MAP. In teamwork-oriented approaches, a diagnosis must explain coordination failures: symptoms are disagreements among the agents, and diagnoses are inconsistent beliefs that agents hold about the same objects. In action-oriented approaches, a diagnosis must explain why the MAP has not reached the expected goals: symptoms are observations that do not match with the expected effects of actions, and diagnoses are subsets of actions in plan that can be qualified as abnormal.

In the rest of the Chapter, we use this initial categorization to guide our discussion: in Section 5 we first present teamwork-oriented approaches paying particular attention to the social diagnosis, then in Section 6 we take into account action-oriented solutions.

5. Teamwork-Oriented Approaches

5.1. TÆMS

TÆMS (Task Analysis, Environmental Modeling and Simulation) is a domain-independent modeling language, initially proposed by Decker and Lesser [19], for representing *organizational information* of a multi-agent system. TÆMS is well-suited to model which roles an agent plays within an organization, and how it interacts with the other agents in the same organization.

A TÆMS model decomposes agents' goals and tasks in a tree-shaped hierarchical representation. In a TÆMS hierarchy one is just interested in modeling how to achieve high-level goals by decomposing them into tasks and subtasks. The resulting structure looks like a tree in which each path terminates with an executable method. Each method is modeled

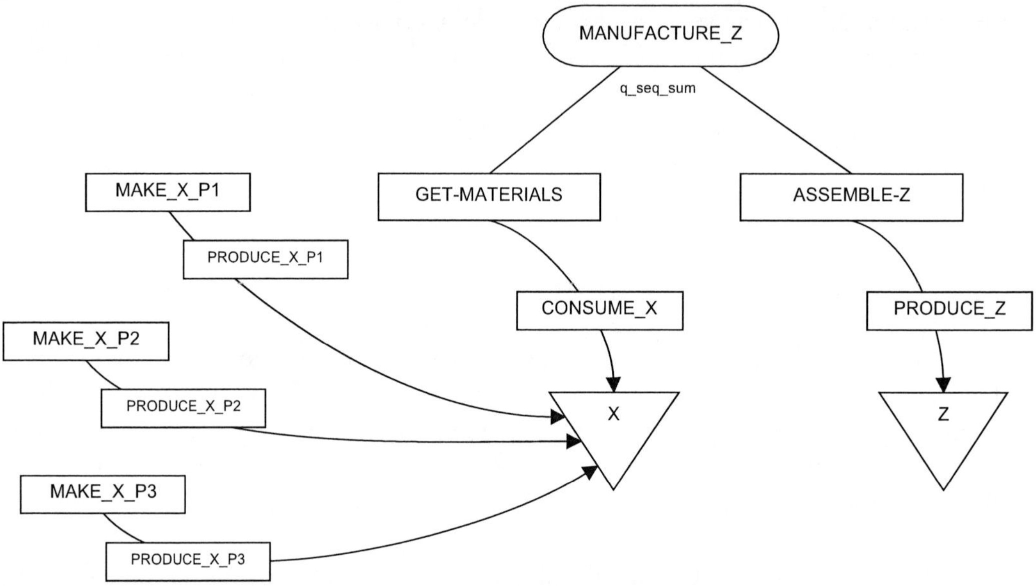

Figure 2. TÆMS task structure example for the production of resource Z.

by specifying its expected (i.e., nominal) behavior, its interactions with other methods and the resources that it will consume. In order to estimate how well a method is performing, TÆMS allows also to associate the method with measures about its expected quality, cost, and duration.

A key factor in TÆMS is the possibility to relate a method (or task) with others by expressing three types of interrelationships:

- *hard*: method A enables/disables method B; the execution of B is therefore conditioned on the execution of A;
- *soft*: method A facilitates/hinders method B; the execution of B is not exclusively conditioned on the execution of A, but other contextual conditions have to be taken into account;
- *resource*: method A consumes/produces a resource r that is in turn produced/consumed by another method B.

Figure 2 shows an example of a TÆMS goal decomposition borrowed from [25]. In particular, the picture shows how the high-level goal of an agent – producing an item of the product Z – is decomposed into different subtasks and methods. It is important to note that some of the resources required by this agent, as for instance X, are provided by other agents in the system (not represented in this picture).

It is easy to see that such a structure is very reach and can be used to generate expectations about the correct behavior of the system.

In [25], Horling et al. focus on one central aspect about MASs: to keep a MAS effective over time, its internal organization must be adapted according to the contextual conditions that arise while the system is working.

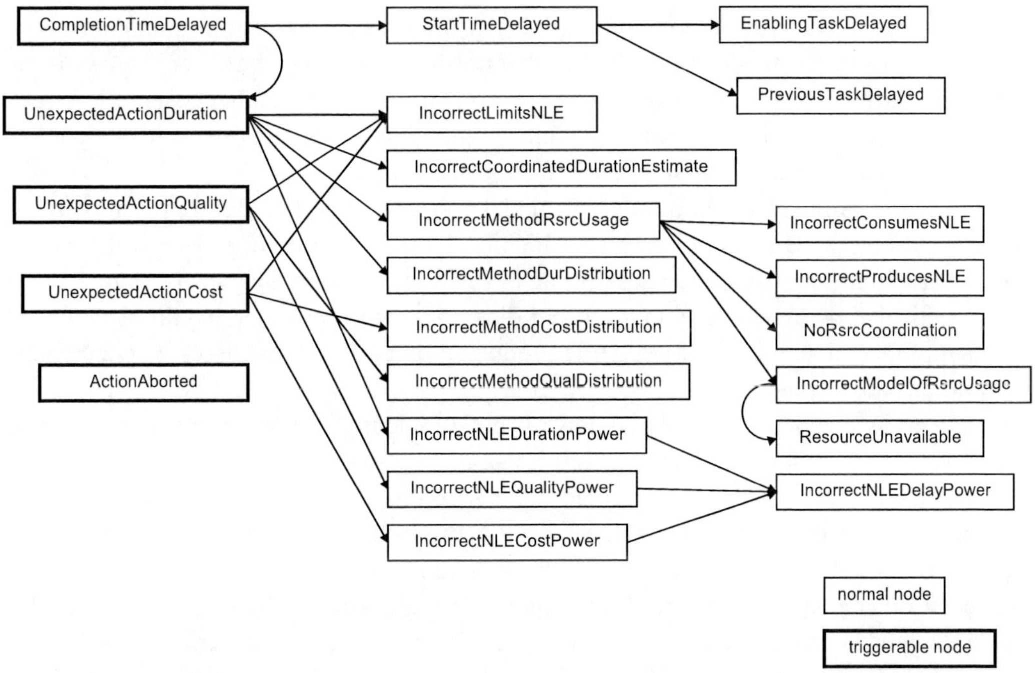

Figure 3. The diagnostic causal model proposed in TÆMS.

To cope with this problem, Horling et al. point out that to properly adapt the organization of a MAS, the adaptation process must be driven by a diagnosis about the behavior of the MAS itself. The idea, in fact, is that the adaptation can intervene more effectively once we have isolated the causes for the observed anomalies. From their point of view, the anomalies to be detected and diagnosed are deficiencies in the way the agents cooperate with one another; these deficiencies typically cause low quality results, low performances, and increased resource consumption.

In the methodology proposed in [25], each agent is in charge of inferring local diagnoses starting from *symptoms* (i.e., interpreted observations). In particular, an agent relies on two components generating symptoms: *observers* and *modelers*. Observers monitor the agent's activities and components and generate symptoms when the observed behavior is not compliant with the one expected according to the TÆMS model. Modelers, on the other hand, are used to adapt the model itself via learning techniques.

Once symptoms are available, the diagnostic process can start. To keep their approach as general and domain-independent as possible, Horling et al. propose the adoption of a causal model; that is, a directed, acyclic graph where nodes are associated with diagnostic problems and with the corresponding diagnostic engines to solve them.

Their idea, in fact, is that to take real advantage of the results in the MBD field, different diagnostic engines should be made available to their diagnostic process. Horling et al. observe that in a MAS there is a wide range of types of faults, but diagnostic engines are typically designed to deal with specific fault types, thus a single diagnostic engine is not sufficient to cover all the possible situations.

Figure 3 shows an example of the proposed causal model. Each node resolves a problem with a specific level of precision and complexity, and hence the resulting diagnoses can be put forward in the model so that other diagnostic nodes can refine them (possibly by exploiting further data coming from the observers). In this way, the diagnosis process can start from easily detectable symptoms explained with high-level diagnoses, which are made more precise along time by passing through the causal model.

It is worth noting that such a causal model can accommodate different kinds of diagnostic processes. Both action and coordination failures can be diagnosed if diagnostic engines are included within the causal model. In addition, some diagnostic engine could just be activated by a symptom and return a diagnosis; whereas others could be activated by a symptom and stay active for a while producing diagnostic hypotheses whenever new observations are available.

As soon as diagnostic hypotheses have been sufficiently refined and their confidence levels pass a given threshold, a reaction generator intervenes by changing the current organization of the agents. For instance, the symptom: "low performance" could be explained by a scarcity of items X that agent A produces and agent B consumes. This explanation could be further refined by individuating that A and B do not coordinate with each other about X. The reaction generator is a specialized agent that, driven by such a diagnosis, imposes a coordination between the two agents over resource X. This is implemented by adding an appropriate relationship within the TÆMS model representing the system.

Possible drawbacks. In some respects, the reaction generator by Horling et al. resembles the institutional agents proposed by Dellarocas et al. [20]. In fact, the reactive generator imposes its decisions on the system as well as the institutional agents resolve contests among citizens. However, Dellarocas et al. have pointed out that the institutional agents have to cooperate in their surveillance task. On the other hand, the solution proposed by Horling et al. appears to be strongly centralized. (Note that, the causal model is about the whole system.) Thus it is not clear how efficiently the approach scales up.

The strength of the diagnostic approach proposed by Horling et al. is its generality. However, it may also hide the main weakness of the approach. In fact, while the causal model leaves the designer free to add, remove, and change the diagnostic engines in order to obtain the best possible configuration for the problem at hand, it also arises the problem of how different diagnostic engines can actually work together.

A first issue is to find a common language that all the diagnostic engines use to exchange information. Since the hypothesis made by a diagnostic engine can be refined by another one, the two diagnostic agents must be able to communicate with each other.

A more important aspect is about the semantic one gives to the dependencies existing among the nodes in the causal model. For instance, when a diagnostic hypothesis inferred in a node can be refined by two or more nodes, is the hypothesis actually refined in only one of the subsequent nodes, or is it refined in all of them? Both scenarios open some critical questions. In the first case, a diagnostic agent has to select, among all its successors in the causal mode, the (best) diagnostic agent to refine its hypotheses. In the second case, when the hypotheses are refined by all the successor diagnostic agents, it may be possible that diagnoses that are locally consistent, are not globally consistent. That is, the refinement inferred by a diagnostic agent may contradict the refinement inferred by another one.

5.2. Social Diagnosis

The term *social diagnosis* has been used the first time by Kalech and Kaminka in [32], and refers to the problem of finding the causes for failures in maintaining social relationships among the agents in a team. Social diagnosis builds up on the seminal works by Tambe [57, 35] about teamwork and coordination failures. However, while in [35] the attention is mainly focused in detecting the teamwork failures, in the works by Kalech and Kaminka the attention is focused on singling out the causes for these failures; that is, in diagnosing them.

Basic Framework. The framework at the basis of the social diagnosis is substantially the one proposed by Tambe and discussed in the previous section. The team of cooperating agents is represented in abstract terms by means of a hierarchy of *behaviors*, which closely resembles the Tambe's hierarchy of team operators. Each behavior is characterized by a set of logic propositions representing the preconditions and the termination conditions of that behavior. A behavior is an abstraction of the concrete actions that an agent actually takes in the world when assumes that behavior.

At any given time of the execution, each agent in the team has a state that is a pair $\langle BP, BL \rangle$; where BP is the current behavior path (within the behavior hierarchy) selected by the agent, and BL is the set of current beliefs of the agents. Agents' beliefs are set of propositions that can either be true or false.

As in [35], the agents synchronize themselves to select a joint team behavior, and as in [35], such a synchronized selection might arise a disagreement when at least two agents have selected two behaviors that are incompatible with each other. Since agents selects their behaviors according to their beliefs, a diagnosis for a disagreement; i.e., a social diagnosis, is a set of conflicting belief states.

In other words, if agent A has selected a different behavior from agent B, there must exist at least one proposition p that A believes true while B believes false. Proposition p is therefore the conflict over which the two agents have to negotiate in order to resolve the disagreement [36].

A consistency-based definition of social diagnosis in MBD terms is given in [29]. This definition identifies a consistency-based coordination diagnosis as a minimal subset of agents in the team that can be qualified as abnormal. In other words, the abnormal agents have selected a behavior (or action) that is conflicting with the ones selected by all the other teammates. By changing the "minds" of the agents qualified as abnormal, one can push them to select different behaviors and hence restore the coordination in the team.

Two-phase Approaches. Social diagnosis has been developed by Kalech and Kaminka in a number of works (see e.g., [28, 27, 30]). It is resolved in two phases:

- *Agent-Selection*: selection of the agent(s) in charge of carrying out the diagnosis;
- *Disambiguation*: how to diagnose a team state.

The approaches mentioned above differ one another in the way the two phases are actually resolved. Both phases are critical as they have a significant impact on the amount of messages that must be exchanged among the agents, and hence on the overhead to be paid for inferring a diagnosis.

In [32, 27] some alternative basic solutions for the two phases are proposed. The problem of selecting the diagnostic agent(s) can be resolved in three different ways:

1. *Preselected*: the diagnostic agents are selected at design-time.

2. *Detectors*: the diagnostic agent is the one that detects a disagreement; since a disagreement can be detected by many agents, each of them will start a diagnostic process on its own.

3. *Estimated Optimal*: the diagnostic agent is the one that reduces the amount of messages exchanged among the agents.

The first solution is the simplest one, but it requires that whenever a failure is detected, it must be declared (i.e., broadcast) in the team so that the diagnostic agents start their inferences. This solution therefore requires a huge amount of messages to be exchanged among the agents.

To cope with this problem, in the strategy based on detectors the agent that discovers a disagreement is the one that starts a diagnostic process. However, the agent that detects the disagreement might not necessarily be the best one to diagnose it. In fact, to diagnose a disagreement, an agent may need to ask the other agents for details about their beliefs in order to build a global context and infer a diagnosis.

The third proposed strategy tries to overcome this inconvenient by selecting the agent that is (expected to be) the best one. That is, the agent that needs to acquire from others the smallest amount of knowledge, and hence needs to exchange the least number of messages. In this case, the diagnostic agent is selected on-the-fly in a distributed way. Each agent in the team runs the same algorithm used for estimating the beliefs held by the other team members. On the basis of the estimated beliefs, it is possible to determine which agent is expected to make least queries to disambiguate the diagnosis. Thus all the agents select the same diagnostic agent without the need to communicate with one another.

The second phase of social diagnosis, the solution of a diagnostic problem, can be resolved in two main ways:

1. *Reporting*: this is the simplest way to resolve a diagnostic problem. Each agent provides the diagnostic agent(s) with its beliefs about its current behavior. Thus, only the relevant beliefs are exchanged. The disagreement in fact is about the current behaviors selected by the agents, so if a belief does not appear in the preconditions of the current behavior, that belief cannot be blamed for the disagreement. The receiving diagnostic agent(s) can therefore compare the beliefs from an agent with the beliefs from the others, and detect the conflicting beliefs.

2. *Querying*: this technique tries to reduce the number of messages that the agents need to exchange with one another. The querying algorithm can be divided into three main steps:

    (a) the diagnostic agent(s) monitors the other agents in the team (see the techniques discussed in [35, 34]), in order to guess their current behaviors;

(b) based on the hypothesized agent behaviors, the diagnostic agent(s) infers the possible beliefs of the other agents, and hence builds an initial set of conflicting beliefs (i.e., diagnostic hypotheses);

(c) to disambiguate diagnostic hypotheses, the diagnostic agent(s) can make targeted queries to limit the number of exchanged messages.

By combining the strategies for the *Agent-Selection* phase with the strategies for the *Disambiguation* phase, Kalech and Kaminka have discussed four different methodologies as reported in Table 1

Table 1. Solving social diagnosis: alternative methods

Method	Agent-Selection	Disambiguation
1	all agents preselected	reporting
2	subset of k detectors	querying
3	one agent preselected	reporting
4	one agent estimated optimal	querying

From their experimental analysis resulted that the runtime is strongly affected by the disambiguation strategy adopted in the *Disambiguation* phase. In particular, methods based on reporting results to be more efficient as the diagnostic agents do not need to reason about the beliefs held by the other agents in the team. However, these methods suffer from a significant communication overhead due to the amount of messages that must be exchanged. This overhead is substantially reduced in methods adopting centralized diagnostic algorithms combined with targeted queries.

Improving the Querying Strategy. Kalech and Kaminka have further improved their querying method. In [28, 30] they have proposed three strategies to make their querying methodology effective when the team involves a very large number of agents. The strategies they propose can be summarized as follows:

- *Behavior Querying*: rather than trying to infer the behaviors of the other agents, the diagnostic agent(s) can ask them what behavior they have currently selected. This solution avoid the diagnostic agent to run complex behavior recognition techniques that are very computational expensive. Moreover, since behavior recognition techniques just rely on observations about the agents' activities, and since observations might not be complete, the behavior recognition process can just infer a set of behavior hypotheses that must be subsequently refined by the diagnostic agent. By means of the behavior querying method, these complexities are avoided.
- *Shared Beliefs:* asking the behavior to each agent in the team might not be always possible or convenient. The number of messages to be exchanged is, in fact, at least linear with the number of agents in the team, but it is more than linear when there

are many diagnostic agents. Shared belief relies on a light-weight behavior recognition technique that allows the diagnostic agent to identify abstract conflicting beliefs. More precisely, while the basic querying technique requires the diagnostic agent to infer conflicting beliefs, the shared beliefs method requires the diagnostic agent just to isolate which propositions are not agreed by at least two agents in the team. In other words, in the basic approach, the diagnostic agent has to precisely infer the beliefs of each team agent, and then infer the conflicting belief. For instance, the diagnostic agent might conclude that agent A believes proposition p is true, while agent B believes p is false. With the shared belief technique, it is not necessary to be so precise. The idea is that there is a disagreement between agents A and B about p, and this suffices to start a negotiation process between A and B.

- *Grouping:* in this case the aim is to reduce the number of exchanged messages by dividing the agents into groups. All the agents that have selected the same behavior are placed in the same group. Of course, the diagnostic agent can either ask (behavior querying) or infer (shared belief) the behavior of each agent, and then place the agent in the appropriate group. The idea is that all the agents in the same group are in agreement (i.e., share the same beliefs). On the other hand, a disagreement about at least one proposition exists between any two agents from two different groups. At this point, in order to diagnose the entire team of agents, it is sufficient to consider a representative agent selected from each group. The diagnostic agent can therefore isolate the conflicting beliefs between any pair of representative agents in order to infer a social diagnosis at global level.

  One possible disadvantage of *grouping*, as the authors themselves state, is that it relies on the assumption that agents within the same group have the same beliefs. This assumption might not hold as agents are grouped on the basis of their behaviors. But behaviors are selected by an agent according to their current beliefs and to the preconditions associated with the behavior itself. It may therefore be possible that agent A believes p, while agent B believes q, but both select the same behavior because it is either enabled by p or q. Kalech and Kaminka do not discuss an effective solution to this potential problem. Indeed, since most of the behavior-selection steps are synchronized among all the team members, they guess that it is very unlikely that two agents select the same behavior having different beliefs, and in their extensive experimental studies they never observed such a condition.

Possible Drawbacks. The strategies about social diagnosis presented so far, however, rely on some assumptions that might not be applicable in many real-world situations. First of all, these strategies assume that the agents' beliefs are homogeneous. Moreover, agents must belong to the same organization as they must share a behavior hierarchy and select, in a synchronous way, behaviors within that hierarchy. On the one hand, these assumptions allow the agents to communicate with one another by exchanging their beliefs. In fact, agents' beliefs can just mentions propositions appearing in the preconditions and post conditions of the behaviors in the hierarchy. On the other hand, the assumptions limit the freedom of agents: (1) agents must be all the same (no specialized behaviors nor beliefs), and (2) agents cannot join and/or leave the team as in the Internet agents considered by Dellarocas in [20].

Moreover, agents must be willing to exchange their own beliefs, or must be aware that other agents are going to monitor their activities. When agents are not willing to communicate their own beliefs to the diagnostic agent, the diagnostic agent is forced to monitor the behavior of all the team members. However, this might not be sufficient when the agents have not a common language to represent their beliefs. In fact the diagnostic agent could not infer conflicting beliefs as each agent has its own set of propositions to represents its internal beliefs, and this language might even been not known to the diagnostic agent.

In such scenarios, a more effective way to get a global picture about the changes happening in the system consists in exchanging observations about the environment (i.e., about the shared resources). For instance, in [22, 49, 50, 52] diagnosis is inferred in a distributed way by allowing the agents to exchange observations about the connections between subsystems.

Another characteristic that might arise issues in some applicative domains is that the social diagnosis is solved in a centralized manner. This seems to be very unpractical especially when multiple agents, spatially distributed, operate simultaneously (see e.g., telecommunication networks in [47]). In this scenario, the synchronized selection of behaviors might be difficult to obtain as latencies in the communication might not be negligible. At the same time, the centralized diagnostic agent might be flooded with messages about the agents' beliefs that would make the inference of a diagnosis highly expensive. Indeed, although the grouping method in [28, 30] has been designed to deal with large-scale teams, this solution is effective only when the agents can be grouped in few clusters, otherwise there is not a substantial benefit.

CSP-based Approaches. Social diagnosis has also been solved with alternative methodologies. In [31], Kalech et al. present an approach to social diagnosis based on distributed Constraint Satisfaction Problems (CSPs). In this case, the diagnostic framework models each agent with a single variable whose domain is the set of actions the agent can perform. Coordination among the team agents is modeled via a set of constraints. Two types of constraints are considered: (1) concurrency constraints impose that two (or more) agents perform specific actions simultaneously; (2) non-concurrency constraints impose that two (or more) agents never try to perform particular actions simultaneously. These two kinds of constraints are used to define which joint action selections are acceptable, and which others are forbidden.

Also in this alternative diagnostic framework, actions are assumed to be fault-less; failures in the systems are only due to the erroneous selection of the actions the agents have to perform. Specifically, a joint selection of actions is erroneous when it violates at least one constraint.

The social diagnosis is defined as a minimal set of agents responsible for having selected the wrong actions. In fact, by changing the selection made by these agents, the coordination among the agents is made consistent again.

In [31], Kalech et al. present four algorithms for inferring such a notion of social diagnosis. One of these algorithms is based on the synchronous backtracking algorithm ([60, 61]), and guarantees to find all the minimal diagnoses. In particular, this algorithm requires the agents to explore the entire space of valid joint action selections. To do so, agents are organized in a ring: each agent sends to its two neighbors the set of actions it can take, and receives positive or negative (i.e., *nogood*) feedback. However, since this algorithm is expensive, the authors propose other solutions; for instance, rather than exploring the whole space, the agents stop their inferences when they find the first possible diagnosis. Of course, this algorithm is faster than the previous one, but it does not guarantee that the inferred diagnosis is minimal.

An important characteristic of all the four algorithms discussed in [31] is that, relying on the exchange of messages, each agent in the team is able to infer a diagnosis for the whole system. Thus, as in the two-phase framework, agents' states are not private but fully accessible by any other team member. This could be a limit in those scenarios where agents' states should be kept private.

Matrix-based Approaches. In [33, 26], Kalech proposes a further alternative framework to solve a social diagnosis problem. In this case, a formalism based on matrices and operations among matrices is adopted. The basic idea is that the admissible combinations of agents' states and actions that might occur in the system can be modeled as a Boolean coordination matrix C. This matrix has as many rows as there are agents in the team, and as many columns as the number of possible actions. The coordination matrix C models the capabilities of each agent: if element $c_{i,j} \in C$ is 1 (i.e., true) agent i can perform action j, otherwise $c_{i,j}$ is 0 (i.e., false) meaning that agent i is not allowed to perform action j.

At the same time, observations about the agents' activities (i.e., which actions they perform at each time instant), can be modeled as a matrix Ω that has a similar structure to matrix C. Matrix Ω represents the actions actually taken by the agents, so $\omega_{i,j} \in \Omega$ is 1 iff agent i is performing action j, false otherwise. Of course, since an agent can just perform one action at a time, each row of matrix Ω contains just a single '1'.

To detect a coordination failure (i.e., a joint selection of actions that violate the coordination constraints), it is sufficient to combine the two matrices: R=$C \wedge \Omega$. If the resulting matrix R contains at least one all-zero row, this is a sufficient condition to conclude that a coordination failure has occurred. The agents corresponding to the all-zero rows are considered faulty. A social diagnosis in terms of erroneous agents' states can be inferred by means of an interpretation matrix that binds the actions selected by the agents with the states that have enabled those actions.

Unfortunately, the presence of all-zero rows in matrix R is just a sufficient conditions. It might in fact happen that even though matrix R does not contain all-zero rows, a coordination failure has occurred the same. Thus the matrix-based algorithm proposed by Kalech is not complete.

The matrix-based approach is sufficiently powerful to deal with non-homogeneous agents (i.e., agents with different set of states), partial observations about the actions taken by the agents (i.e., not all the actions are observed precisely), and complex coordination constraints (i.e., coordination requirements modeled via more than one coordination matrices).

The main advantage of this approach is that its computational complexity is polynomial as it just involves operations on matrices. The experimental analysis discussed in [26] prove the effectiveness of the approach; however, the paper does not provide details about its implementation. It is easy to see that the scalability of the approach strongly depends on how good matrices and operations between matrices are handled. A similar problem also occurs in the approaches proposed by Micalizio et al. that we discuss in the next section. To cope with very huge relations, Micalizio et al. have proposed the adoption of the symbolic formalism of the Ordered Binary Decision Diagrams (OBDDs) [6, 7]. It would be interesting to assess whether OBDDs could give a significant contribution in enhancing the efficiency and the scalability of the matrix-based approach proposed by Kalech.

6. Action-Oriented Approaches

In this section we review some approaches to the diagnosis of MAPs in which the plan is explicitly modeled. The basic idea of these approaches is that a given MAP can be thought of as a component-oriented system to be diagnosed: agents' actions are viewed as the components of the system, whereas the causal dependencies between the actions are viewed as the connections between components. Thus, MBD techniques that have been developed for component-oriented systems can be adapted and applied to diagnose a MAP.

Generally speaking, the aim of the diagnostic task, in this case, is to qualify a minimal subset of actions as abnormal since they failed to get the expected effects. The main difference with social diagnosis is about the focus of the diagnosis. In action-oriented approaches, the diagnosis is focused on the plan actions, that can fail for a number of reasons. The diagnosis should be able not only to isolate the faulty actions, but also to identify the root causes of their failures.

In this section we identify three main families of approaches:

- *Automaton-Based Action Models*: actions are durative and modeled as automata. Unexpected, harmful interactions among agents can arise during the execution due to the occurrence of faults in the functionalities of the agents.

- *Resource-Oriented View*: actions are atomic, represented just in terms of nominal preconditions and effects. Agents are not explicitly modeled, and the status of the system is the status of the resources shared by the agents. Observations are about the state of shared resources.

- *Relation-Based Action Models*: actions are atomic, but their models include both the nominal (expected) behavior, and the set of abnormal evolutions that might happen during the execution. Action models are relations over the agents' status variables, thus agents are modeled explicitly. Dependencies between agents' actions are also made explicit in terms of causal and precedence links.

In the rest of this section we review the main approaches falling into these three categories.

6.1. Automaton-Based Action Models

A first attempt to explicitly handle actions in diagnosing MAPs has been proposed by Micalizio et al. in [43], and subsequently extended in [45, 39]. A key characteristic of these works is that they consider durative actions; i.e., actions that last for a while. These actions are modeled as automata that encode both nominal and anomalous action progresses.

Control Loop Architecture. Although in their works Micalizio et al. mainly focus on the diagnostic task, they also provide an insight on how the diagnosis should be integrated within a global control loop that supervises the activities carried out by the agents.

More precisely, in the architecture they propose, a single Planning Agent synthesizes a MAP that gets a human-defined goal. The Planning Agent decomposes the global MAP into subplans so that each agent in the team is assigned with a single sequence of actions to be performed. Indeed, the Planning Agent is also a scheduler as the actions dispatched to the agents are qualified with their expected starting time and duration.

During the actual execution of the plan, a Supervisor Agent monitors the progress of the actions and detects anomalies (e.g., actions that fail to achieve their expected effects within a time deadline). Whenever an anomaly is detected a diagnostic process is activated in order to explain them. The inferred explanations are therefore sent back to the Planning Agent that can synthesize a new (global) plan for reaching the goal taking into account the information provided by the Supervisor Agent.

Note that agents operate simultaneously sharing the resources available in their environment. Observations are made available to the Supervisor Agent by a network of sensors distributed in the environment. Although sensors are assumed reliable, they can detected just a subset of events, thus the activities performed by the agents are just partially observable.

Agents are cooperative as they all concur to obtain the global user-designed goal. Differently from the approaches by Tambe and Kalech about teamwork, however, these agents do not need to maintain a synchronized behavior over time. Some form of coordination among the agents is still required, though, as they can only access resources in mutual exclusion. In fact, when two or more agents need the same resource in the same time interval, they have to negotiate the order in which they will access that resource. Note that agents are autonomous in the negotiation process. In addition, agents are not required to provide the Supervisor Agent with information about their beliefs, nor about the results of a negotiation phase. In other words, the Supervisor Agent relies only on sensors' messages.

Diagnostic Framework. A novelty of this approach, compared with the teamwork approaches, is that the execution of a MAP can fail not only due to *troublesome interactions* among the agents, but also due to *faults* in the agents' functionalities. A troublesome interaction is somehow similar to a coordination failure in teamwork approaches. It is in fact a situation in which two or more agents hinder one another; for example, the competition for a resource is a form of troublesome interaction.

It is important to note that, although in principle the Planning Agent can schedule the actions in a way that limits the number of negotiations among the agents, troublesome interactions are not completely avoidable nor predictable as they might be a consequence of faults. It may happen, for instance, that a robotic agent breaks down in a doorway preventing others from walking through that door.

In the framework proposed by Micalizio et al., the state of each agent i is explicitly modeled as a set END^i of *endogenous* variables representing, for instance, its position, the objects it is currently transporting, its current action, and so on. Besides these endogenous variables, the agent's states also includes a set HLT^i of *health state variables*. Variables in HLT^i model the health state of the functionalities of agent i.

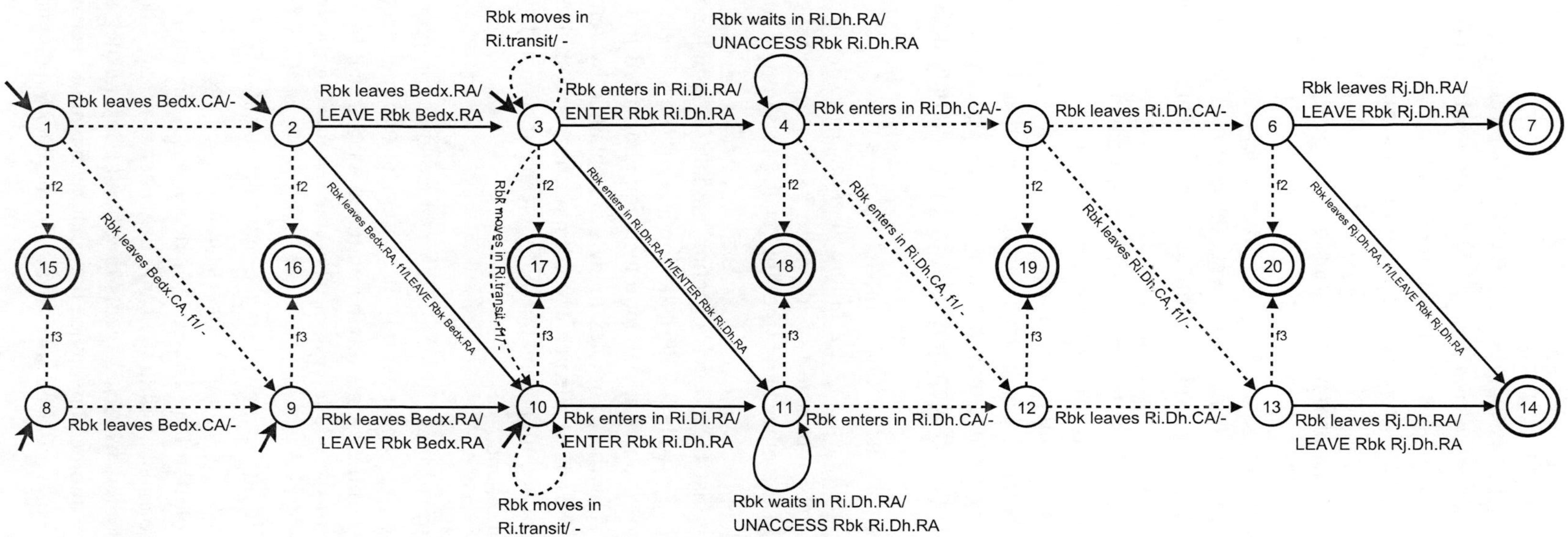

Figure 4. The automaton for the $GoTo(Rb_k, R_i, R_j, D_h)$ action in the *Robo*Care domain.

Note that models of the functionalities establish a relation between action failures and faults. Functionalities are in fact essential to complete actions, and their health conditions have a strong impact on the actions' outcomes. Thus it is important to trace the health of functionalities over time in order to create expectations about the future evolutions of the performed actions.

Expectations are built by means of action models. As anticipated, actions carried on by the agents are represented as automata in order to model not only the initial and final states, but also intermediate situations that might occur during the action execution.[2]

An example of action model is given in Figure 4, showing the *GoTo* action model in the *Robo*Care applicative domain (see [9]). Each node in this automaton represents an agent state, transitions between two states model how the status of an agent evolves over time. In particular, solid edges represent observable state transitions; that is, transitions detected by some sensors; whereas dotted edges are non-observable state transitions.

Note that automata are particularly useful to take into account troublesome interactions without the need to explicitly model them. For instance, to represent that an agent could be involved in a traffic interaction, and hence may wait a certain time before accessing a resource, it is sufficient to introduce a self-loop transition in an appropriate node of the *GoTo* model.

In addition, it is worth observing that such a kind of model specifies which agent's states are either a successful completion of the action, or a failure. For instance, the final states 7 and 14 of the automaton in Figure 4 represent two alternative states for the successful completion of the *GoTo* action. On the other hand, the final states 15 through 20 are all situations in which the action is failed.

In [43], Micalizio et al. show that these kind of action models, although complex, can be obtained by composing (i.e., synchronizing) the models of simpler entities, as agent's functionalities, sensors, and so on. They adopt a technique borrowed from DESs approaches (see for instance [56, 47]) where complex models are obtained by composing elementary blocks. By modeling each of elementary components as communicating automata [4], the *GoTo* action model is built automatically by means of standard synchronization between communicating automata.

Centralized *vs.* Decentralized Approaches. In [43, 45], Micalizio et al. propose a centralized approach to implement the Supervisor Agent. It carries out two main activities: (1) *monitoring* on-line the activities of the agents relying on the messages sent by the sensors, and (2) *diagnosing* the status of the whole system when the monitoring discovers an erroneous situation.

Such a centralized approach, although simple, has the main disadvantage to scale up poorly as the number of agents grows. To cope with this issue, Micalizio and Torasso have proposed in [39] a decentralized solution that dynamically decomposes the team in subsets, called *dependency sets*: all the agents within the same dependency set are diagnosed together as they depend on one another; whereas, agents in different dependency sets are guaranteed to be independent of one another, and hence they can be diagnosed separately.

In the rest of this subsection, we first describe how the monitoring and the diagnosis are

[2]From the point of view of the Planning Agent, durative actions are just modeled in terms of their *nominal* preconditions, effects, and possibly invariant conditions as proposed since PDDL 2.1 [21].

performed in the centralized approach, and then we briefly sketch how these two activities are tailored for subsets of agents.

Centralized On-line Monitoring. The goal of the centralized monitoring task is to infer the state of the whole system at each time of the execution. As a general principle, to reach this goal the monitoring needs two pieces of information: (1) the system state $\mathcal{S}_t$ at time t, and (2) a transition relation Δ_t that models how the state at time t evolves in a state at time t+1.

- *System State and Belief States.* In the framework proposed in [43, 45], the state $\mathcal{S}_t$ of the whole system is set of all the agents' states at time t. However, since the system is just partially observable, the monitoring process is unable to precisely estimate the agents' states; it rather estimates *belief states*, i.e., set of alternative agents' states that are all possible because they are all consistent with the available observations. The centralized monitoring therefore deals with global belief states $\mathcal{B}_t$ rather than global system states. This makes the monitoring process very complex as it has to cope with uncertain information. One of the limits of the centralized approach, in fact, is that to model uncertain information it might build belief states mentioning a huge number of alternative systems states. (In some experimental settings, belief states with more than 100,000 have been observed, see [39, 45].)
- *Transition Relation.* The global transition relation Δ_t consists of the set of action models $\Delta(act_t^i)$ for the actions that are active (i.e., in progress) at time t. More precisely, if agent i is performing action act_t^i at time t, $\Delta(act_t^i)$ is the transition relation of the automaton modeling the action act^i itself.

Having these two pieces of information, the estimation of the belief state about the whole system at the next time t+1 can be formalized in terms of Relational operators as follows:

$$\mathcal{B}_{t+1} = \text{PROJECT}_{t+1}(\mathcal{B}_t \text{ JOIN } \Delta_t) \quad (1)$$

The intuition is that the JOIN operator is the predictive step through which one can estimate the set of possible evolutions in the next time, and then the PROJECT operator projects the estimates over the state variables at time t+1 (see [45] for details).

Note that the belief state $\mathcal{B}_{t+1}$ resulting from formula 1 may contain spurious states that are pruned off by asserting global constraints, especially about the use of resources. For instance, since the troublesome interactions are not explicitly modeled within the action models, it may be possible that in some states estimated by formula (1) two or more agents have access to the same critical resource simultaneously. Of course, these states are not actually possible, and hence domain-dependent constraints are imposed on $\mathcal{B}_{t+1}$ to discard them. Similarly, the belief state $\mathcal{B}_{t+1}$ is also refined by exploiting the set of observations coming from the sensors: all the states which are inconsistent with the sensors' messages are pruned off.

The centralized monitoring detects anomalies in two ways. First, it may happen that the state estimated in $\mathcal{B}_{t+1}$ for agent i corresponds to one of the failure states in the model of action act_t^i currently in progress. This anomalous situation must be explained by the diagnostic module in terms of faults in some functionalities of agent i.

Second, an action could run longer than expected according to the schedule given by the Planning Agent. In particular, whenever a new action starts its execution, the monitoring process starts two timers: $\tau_{delayed}$ and τ_{failed}. The first timer is set to the expected ending time according to the given schedule. When an action terminates successfully after that its corresponding $\tau_{delayed}$ timer has expired, the action is labeled as *completed with delay*.

The second timer τ_{failed} is introduced to accommodate the delay that an agent could gather due to unexpected troublesome interactions. This timer is set to the maximum time the Supervisor is disposed to wait for the completion of the action even though with some delay. If an action is not completed when its τ_{failed} timer expires, the action is labeled as *failed*.

Both situations, actions completed with delay or actions not completed at all, are anomalies that the monitoring detects and passes them forward to the diagnostic procedure.

Centralized Diagnosis. The diagnostic task is resolved by exploiting a base of causal rules that map anomalous situations with their possible causes. These rules have the following general form:

$$cause_1 \wedge cause_2 \wedge \ldots \wedge cause_n \rightarrow anomaly_1 \vee anomaly_2 \ldots \vee anomaly_m \quad (2)$$

These rules are used abductively starting from the anomalies detected by the monitoring. For example, let us assume that the monitoring has detected that the *GoTo* action performed by agent i has outcome *delayed.* The diagnostic module retrieves from its knowledge base all the rules that can be used to explain the outcome of a *GoTo* action, and selects all and only the rules that mention the anomaly *delayed* in the consequent. In the *Robo*Care domain these rules are:

$$i.mob(ok) \wedge D_h.traffic(moderate) \wedge D_h.occl(no) \rightarrow outcome(on\text{–}time) \vee outcome(delayed) \quad (3)$$

$$i.mob(slowdown) \wedge D_h.traffic(no) \wedge D_h.occl(no) \rightarrow outcome(delayed) \quad (4)$$

$$i.mob(slowdown) \wedge D_h.traffic(moderate) \wedge D_h.occl(no) \rightarrow outcome(delayed) \vee outcome(failed) \quad (5)$$

Rule 3 means that a *GoTo* action can be delayed by a traffic interaction occurring while agent i attempts to walk through door D_h. Rule 4 means that the same delayed outcome can be obtained as a consequence of a fault in the mobility functionality: the mobility is no longer working in its nominal mode ok but in the $slowdown$ mode. Finally, rule 5 means that a delayed action can be a consequence of a combination of a fault in the mobility functionality and of the traffic interaction.

Although these three rules could be provided as possible (abductive) explanations, it is possible to infer more precise explanations by assessing whether the traffic interaction has really occurred or not. To do this it may be possible to inspect a history of belief states and assess whether some necessary of sufficient conditions for the presence of traffic are satisfied; this refinement technique has been investigated in [44].

Decentralized Monitoring and Diagnosis. As mentioned above, the centralized approach does not scale up well when the number of agents in the team grows. To cope with this issue, Micalizio and Torasso have proposed in [39] a decentralized architecture in which a team of *robotic agents* is supervised by a team of *supervisor agents*: each robotic agent is monitored and diagnosed by a specific supervisor agent. The idea is that each supervisor agent receives observations about the robot it is associated with, and relying on these observations it has to detect robot's misbehaviors (i.e., delays or failures of the actions assigned to the robot). Each supervisor agent, however, cannot work isolated from other supervisor agents. In fact, since robots may compete/negotiate for accessing the resources, the supervisor agents have to cooperate with one another in order to acquire an amount of information sufficient to understand how well the monitored robots are operating.

The principle of the decentralized solution is that, in general, a supervisor agent does not need to interact with all the other supervisor agents, but just with a subset; i.e., a *dependency set*. More precisely, whenever two or more robots are performing actions that require the same resources (these robots may compete for accessing the resources), the corresponding supervisor agents are put in the same dependency set. At each time of the plan execution, only the agents within the same dependency set need to cooperate with one another. Of course, dependency set are dynamically detected as they depend on the actions that are active at any time instant.

Cooperation among supervisor agents is achieved via the exchange of high-level information. Indeed, agents within a dependency set exchange one another intermediate results. These results come from inferences the agents make by exploiting the direct observations about the monitored robots. The advantage is that the amount of data to be exchanged is reduced because partial results are more synthetic than the original messages sent from sensors. In addition, this guarantees some form of privacy as the direct info about the monitored robots is hidden, and only synthetic (high-level) information is actually exchanged.

Once each agent within a dependency set has received a piece of information from the other members of the dependency set, it can build a partial system state (i.e., a *dependency-set belief*) that is guaranteed to be globally consistent. On the basis of such a local structure, each agent within a dependency set can detect (in isolation) anomalies about the monitored robot, and consequently activate a diagnostic process to explain them. Thus a global diagnosis is not built, but it could be obtained by combining the local diagnoses inferred by the agents.

Advantages and Disadvantages. It is easy to see that the decentralized solution scales up better than the centralized one. In fact, the centralized solution has to build a global belief state for the whole system creating dependencies between agents that are actually independent of each other. On the other hand, the decentralized solution creates a belief state for each dependency set active in a given time. These belief states are in general much smaller than the global belief state as they mention only a few robots: In spatially distributed environments, robots may interact with one another only for short time periods when they are nearby a resource. Most of the time, however, a robot does not have to interact with others, and the associated supervisor agent does not too.

Comparing the decentralized solution proposed by Micalizio et al. with the two-phase social diagnosis proposed by Kalech et al. we can draw some interesting results. The decentralized solution allows the supervisor agents to keep some form of privacy about the monitored robots as they exchange one another only synthetic information. This could be useful when the robots in the team are heterogeneous: By exchanging high-level data, each agent needs to know in detail just the robot it monitors, while details about the other robots are hidden. This means that the belief states maintained by each supervisor agent are tailored on the monitored robot. On the other hand, the two-phase social diagnosis implicitly assumes that all the agents are homogeneous in their internal belief states.

In addition, in the decentralized solution the agents within a dependency set cooperate to get a common result: They are all at the same level; each of them infers a local diagnosis that is globally consistent. In the two-phase social diagnosis, instead, the diagnostic agent is the only one that can ask information to others, and it is the only one that infers and knows the diagnosis of the whole team.

A possible disadvantage of the automaton-based action model approach, is that it requires complete fault models. That is, for each action type, the designer has to provide the set of possible faults that may affect the functionalities used during that type of action. For each fault, the designer has to specify how the fault may affect the evolution of the action itself. Moreover, the designer should also individuate which troublesome interactions may affect the outcome of each type of action. This may not be an easy task since, as Tambe [57] has already noted, some kind of troublesome interactions are discovered only when the system is running.

The task of the designer in the two-phase social diagnosis by Kalech et al. is much simpler. Behaviors abstract fine-grained actions; thus different types of erroneous action executions or of erroneous sensor readings are actually modeled as a wrong synchronized selection of a team behavior. The designer, therefore, has just to reason about behaviors and behavior hierarchies, and she/he does not need to model specific fault modes to the actions performed by the agents.

Finally, a possible issue in dealing with relations is that they might have huge sizes. To be effective, the automaton-based approaches must handle relations efficiently. Micalizio et al. have proposed in their works to symbolically encode relations as Ordered Binary Decision Diagrams (OBDDs), [7, 6], so that the Relational operators involved in the prediction process and in the inference of a diagnosis can be efficiently implemented in terms of standard operations between OBDDs, see [58, 38] for more details.

6.2. Resource-Oriented Approaches

Roos and Witteveen propose in many of their works to think of a MAP as a component-based system to be diagnosed. They thus suggest to apply MBD techniques to such a kind of system. In particular, the model of such a system refers to the plan actions and the resources the agents are allowed to use. Actions are viewed as the system components that can be qualified as abnormal when something wrong occurs. Connections between actions are realized by means of the resources. Roughly speaking, actions take in input resources in some states, and give in output resources with different states. Thus, observations are about the resources' states rather than about agents' states as in the automaton-based framework.

Resource-Oriented Framework. Witteveen et al. [59] present the basic framework that they use in many of their subsequent works. We call such a framework *resource-oriented* since the state of the system at a given time is modeled via the states of its resources. More formally, let $Obj = \{o_1, \ldots, o_n\}$ be a set of resources (or objects) such that each resource

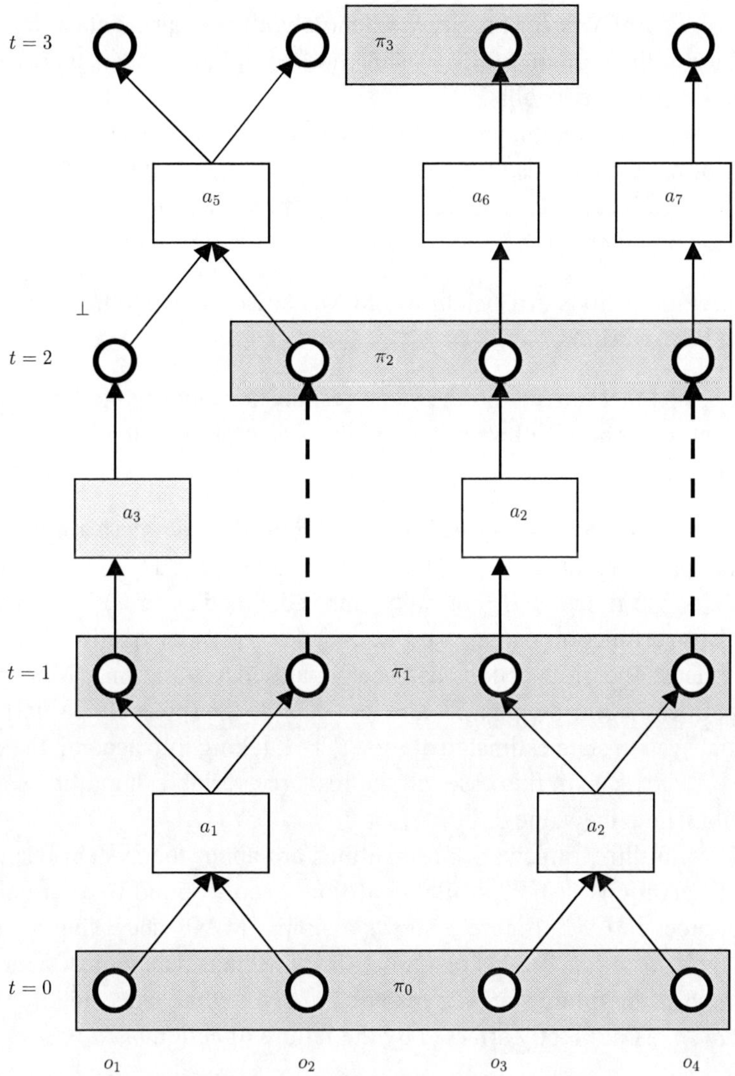

Figure 5. An example of MAP according to the resource-oriented view.

o_i has a finite set of possible states S_i, then the state σ of plan execution at any time belongs to $S_1 \times \ldots \times S_n$. In general, however, only partial states $\pi \in S_{i_1} \times \ldots \times S_{i_k}$, (where $1 \leq i_1 < \ldots < i_k \leq n$) are available. All the resource variables that are not mentioned within a partial state π assume the value $\perp$ meaning that their values are *undefined*, or *unpredictable*.

Actions are modeled as mapping functions between partial states. More precisely, given the action template o (called *plan operator*), its *nominal* model f_o^{nor} is defined as:

$$f_o^{nor} : S_{i_1} \times \ldots \times S_{i_k} \rightarrow S_{j_i} \times \ldots \times S_{j_l}, \quad (6)$$

where $1 \leq i_1 < \ldots < i_k$ and $\{j_1, \ldots, j_l\} \subseteq \{i_1, \ldots, i_k\} \subseteq Obj$.

The abnormal behavior of a plan operator, on the other hand, is very weakly modeled as it is just a function mapping any partial states into a tuple of undefined values; i.e.,

$f_o^{ab}(\pi) = \langle \bot, \dots, \bot \rangle$, where for brevity we denote with π a partial state defined, at least, over the variables in the domain of f_o^{nor}. This model implicitly states that the behavior of an abnormal action is unpredictable.

Plan operators are just templates used to instantiate the actions (called *plan steps*) the agents will perform. A plan P is therefore a tuple $\langle A, Obj, \prec \rangle$, where A is the set of plan steps, Obj is the set of resources, and $\prec$ is a partial order relation between actions.

The execution of a plan must respects two important assumptions:

1. Agents perform actions in a synchronized way and each action takes exactly one time instant to be performed.

2. Two plan steps that operate on the same resources cannot be performed at the same time instant. Roos and Witteveen refer to this assumption as the *concurrency requirement* or as the *determinism assumption*.

Relying on the assumptions above, it is possible to define a derivability function that given a (possibly partial) plan state (π, t) at time t and the plan P, allows us to predict the nominal state (π', t') at time $t' > t$. Formally, this is denoted as $(\pi, t) \rightarrow_P^* (\pi', t')$.

Such a derivative function can also be extended to consider qualified plans, that is, a plan in which at least one of its actions has been qualified as abnormal. More precisely, let $Q \subseteq A$ be a qualification for the plan P, and let (π, t) be the (possibly partial) plan state at time t, the partial system state estimated at time $t' > t$, taking into account the qualification Q, is $(\pi, t) \rightarrow_{Q;P}^* (\pi'_Q, t')$. In this case, all the resource variables modified by at least one action in Q will assume the value $\bot$ in π'_Q.

As said above, in this framework observations are about the system resources. More precisely, the observations $obs(t')$ available at time t' correspond to a partial state about a subset of resources in Obj. Figure 5 shows a simple MAP[3] according to the resource-oriented view by Roos et al. In this example, the abnormal behavior of action a_3 makes impossible to predict the future states of resource o_1. Note that, since action a_5 requires in input resource o_1, a_5 is indirectly affected by the failure of action a_3.

Diagnostic Window. A diagnostic problem is defined within a time window $[t, t']$; at time t, we have a snapshot of the plan state (π, t), which can also be just partial; at time t' we have another snapshot of the plan state given by the observations $obs(t')$. A diagnostic problem arises when the partial state in $obs(t')$ is inconsistent with the partial state (π', t') derived from (π, t) assuming that all the actions performed in $[t, t']$ (denoted as $P_{[t,t']}$) behave normally. Note that $obs(t')$ and (π', t') are inconsistent when there exists at least one variable v such that, v is mentioned in both partial states, but the value of v observed in $obs(t')$ differs from the value of v estimated in (π', t'). To simplify the discussion in the rest of this section, we will give for granted the existence of a specific diagnostic window $[t, t']$ over which the diagnostic must be solved.

The solution to a diagnostic problem is a qualification $Q \subseteq A$ such that, assuming the actions in Q are abnormal, allows us to estimate a partial state (π'_Q, t') *compatible* with the observations $obs(t')$. Two partial states are compatible when for each variable $v \in Obj$

[3] This example is a simplified version of an example appeared in [59].

mentioned by both states, v assumes the same value in the two states, or is undefined in at least one of them.

Relying on such a basic framework, Roos and Witteveen have explored the plan diagnosis problem from three main perspectives:

1. Plan disruptions are due to actions, and actions fail independently of one another.
2. Plan disruptions are not only due to actions but also to synchronization errors.
3. Plan disruptions are due to actions, but actions failures are not independent of one another.

In the rest of this section we briefly sketch these three different extensions to the basic framework. Then we briefly sketch a further extension to the resource-oriented framework where the plan states encompasses, besides resources, also other entities such as agents and equipment.

Actions Failing Independently of One Another. This scenario has been mainly addressed in [55]. The key point is that defining a plan diagnosis as a subset of actions that can be qualified as abnormal is, in many cases, insufficient. Since the number of diagnoses is potentially huge (exponential in the number of plan steps), one is interested in diagnoses that are better than others according to some preference criteria. In classical MBD it is often assumed that failures are independent of one another; this implies that diagnoses that assume the minimal number of faults are preferable. According to this principle, Roos and Witteveen propose to prefer plan qualifications that are subset-minimal, and especially minimum cardinality diagnoses. More precisely, the two definitions below follow directly from the MBD literature [24].

Definition 6. *A qualification Q is a* subset minimal plan diagnosis *if for any other plan diagnosis Q', such that $Q' \subseteq Q$, then it holds that Q' equals Q.*

Definition 7. *A qualification Q is a* minimum cardinality plan diagnosis *if for any other plan diagnosis Q', it holds that $|Q| \leq |Q'|$.*

Roos and Witteveen, however, make a step further. They in fact observe that different minimal plan diagnoses (or minimum cardinality plan diagnoses) have different *predictive powers*; that is, the predictions they allow us to do can have different precision. Thus it is reasonable to look for diagnoses that are more precise than others. To reach this goal, we need to introduce a further assumption: the principle of *likelihood of failure propagation*. According to this second principle, the likelihood that an abnormal plan step produces (either directly or indirectly) correct results is reasonably low. In other words, it is not possible, or very unlikely, that the effects of an abnormal action are masked by some other (subsequent) actions in the plan. This principle allows us to define the predictive power V_Q of a qualification Q as the subset of variables that are predicted correctly assuming Q itself. Looking for more precise diagnoses, one should look for qualifications that maximize their predictive power: Having two qualifications Q and Q', we would prefer Q if $|V_Q| > |V'_Q|$.

By extending the two definitions 6 and 7, Roos and Witteveen introduce the following notions:

Definition 8. *A qualification Q is* maximally informative plan diagnosis *(maxi-diagnosis) if for any other qualification Q' it holds $|V_Q| \geq |V'_Q|$.*

Definition 9. *A qualification Q is* minimal maximally informative plan diagnosis *(mini-maxi-diagnosis), if (i) Q is a maxi-diagnosis, and (ii) for any other maxi-diagnosis Q' it holds $|Q| \leq |Q'|$.*

An important property of maxi-diagnoses and mini-maxi-diagnoses is that they can be found in polynomial time, and two ways, one centralized and one distributed, of computing them are discussed in [55].

Abnormal Actions and Synchronization Errors. In [53], Roos and Witteveen try to extend the basic framework to cover not only abnormal actions, but also synchronization errors, which somewhat resemble the coordination failures by Kalech and Kaminka in social diagnosis. More precisely, they consider three types of synchronization errors:

- Duplicated actions;
- Omitted actions;
- Precedence constraints violations.

The first two types may occur when the plans assigned to the agents are redundant. For instance, an action a is included in the plans assigned to two different agents i and j. In principle, action a should be performed by only one agent, but at planning time it is not possible to say which agent, either i or j, will be ready first to perform a. During the plan execution phase, the two agents have to synchronize in order to decide which of them will actually perform a. As a consequence of an erroneous synchronization between them, however, it may happen that a is performed twice (both agents perform it), or even that a is not performed at all since i assumes that j will perform it, and vice versa j assumes that i will be in charge of a.

The third type of erroneous synchronization happens when the execution of action a violates some precedence constraints in $\prec$; that is, a is performed before another action a', but the plan structure specifies that a' should precede (directly or indirectly) a.

Plan diagnosis should detect and identify these situations whenever they occur. To cope with this problem, the derivative function, used to predict future plan states, must be extended in order to capture the synchronization errors. In [53], the proposed extension is based on the notion of plan transformations. The original plan is transformed in order to model either the occurrence of a duplicated action, of an omitted action, or of a violated precedence constraints. The transformed plan is therefore used to predict future states as in the basic framework.

The main issue, here, is that solving a generalized diagnostic problem which has to identify a combination of abnormal actions and synchronization errors is $\mathbf{NP}$-*hard*. In some way, this problem is similar to the one faced with automaton-based action models by Micalizio et al., but Roos and Witteveen fail in giving a solution for the general problem. They just propose approximated solutions to find diagnoses that involve just one type of anomaly.

Causal Diagnosis: Dependent Action Failures. Although maxi-diagnoses and mini-maxi-diagnoses can be inferred in polynomial time, these types of diagnoses are not very useful when there exist dependencies between the failure of different actions. That is, when the abnormal behavior of an action has a cascade effect in the whole plan, and propagates causing the abnormal behavior of other actions.

To deal with causal dependencies between actions' health states, Witteveen et al. [59] propose to capture them by means of a set of *causal rules* of the form:

$$\alpha_1, \ldots, \alpha_k \rightarrow \alpha_{k+1} \tag{7}$$

where $\alpha_1, \ldots, \alpha_{k+1}$ are all plan operators (i.e., action templates). The idea is that whenever a plan qualification Q mentions a plan step a instantiating one of the templates $\alpha_1, \ldots, \alpha_k$, then we can qualify as abnormal any other plan step a' such that:

- a' instantiates α_{k+1}, and
- a' follows a (i.e., $a \prec a'$), and
- a' has been performed in the diagnostic window $[t, t']$, $a' \in P_{[t,t']}$.

More generally, having a *causal theory* Φ including a set of rules as in equation (7), one can characterize any minimum plan diagnosis Q by inferring the causal diagnosis associated with Q.

Definition 10. *Given a minimum plan diagnosis Q for the plan P in the diagnostic window $[t, t']$, Q' is a* causal diagnosis *associated with Q if it is a minimal subset of plan steps in $P_{[t,t']}$ such that $Q \subseteq C_\Phi(Q') \cap P_{[t,t']}$.*

Here $C_\Phi(Q')$ denotes the set of actions that can be qualified as abnormal due to some causal rule in Φ instantiated with respect to a plan step $a \in Q'$.

In other words, Q' is a causal diagnosis associated with Q if it is sufficient to assume Q' to infer Q by means of the causal theory Φ. It is easy to see that Q' always exists. In addition, it always holds that $Q' \subseteq Q$.

Unfortunately, the causal diagnosis Q' might be less informative than the associated diagnosis Q. To provide diagnoses that are a good balance between minimality and informative power, Witteveen et al. introduce the notion of *Pareto minimal causal diagnoses.* Intuitively, Q is a Pareto minimal causal diagnosis iff: (*i.*) Q is at least as informative as any other causal diagnosis Q' assuming fewer abnormal actions than Q (i.e., $|Q'| < |Q|$), or (*ii.*) Q is more informative than any other causal diagnosis Q' assuming as many abnormal actions as Q (i.e., $|Q'| = |Q|$).

Finally, it is important to note that, having the causal theory Φ and a qualification Q, one can also estimate the impact of Q in the rest of plan. In fact, the set of actions in $C_\Phi(Q) \cap P_{[t,|A|]}$ can be considered as an estimate about which goals will be achieved (assuming no other failures occur meanwhile), and which others will not.

Extending the Resource-Oriented Framework. The basic resource-oriented framework allows us to infer a plan diagnosis consisting of a subset of plan steps qualified as abnormal. In some cases, however, one wants to understand why some actions behaved abnormally;

that it, one is interested in finding the root causes for the action failures assumed by the plan diagnosis. In [14], de Jonge et al. address such a problem by introducing the notion of *secondary diagnosis*. Roughly speaking, we could say that a secondary diagnosis is a sort of diagnosis of the primary one: a secondary diagnosis explains the assumptions made by the primary diagnosis.

de Jonge et al. identify three main causes for an abnormal plan step a: the *agent* responsible for a might make errors, the *equipment* used to accomplish a might be faulty, and the *environment* might change unexpectedly and influence negatively the progress of a.

The plan state is therefore represented by a set of variables $\mathcal{V}$, which can be partitioned into four main subsets: $\mathcal{S}$ is the subset of plan steps, $\mathcal{A}$ is for the agents, $\mathcal{E}$ for equipment, and $\mathcal{N}$ for the environment. This final set can include resources, as in the basic framework, and also other contextual conditions (e.g., weather conditions) that might influence the execution of the plan actions. Of course, each variable $v \in \mathcal{V}$ has its own domain D_v, that is assumed discrete and finite.

To infer a secondary diagnosis, it is necessary to extend the model of the plan operators to take into account these four subset of variables. The nominal behavior of a plan operator o is defined as follows:

$$
\begin{aligned}
f_o^{nor} : D_s \times (D_{a_1} \times \ldots \times D_{a_i}) \times (D_{e_1} \times \ldots \times D_{e_j}) \times (D_{n_1} \times \ldots \times D_{n_k}) \rightarrow \\
(D_{a'_1} \times \ldots \times D_{a'_u}) \times (D_{e'_1} \times \ldots \times D_{e'_v}) \times (D_{n'_1} \times \ldots \times D_{n'_w}) \qquad (8)
\end{aligned}
$$

The structure is similar to the normal model of a plan operator in the basic framework, see equation (6). Intuitively,

- $s \in \mathcal{S}$ is a specific plan step instantiating the plan operator o (or in planning terms, o is an action template and s a ground action instantiating o),
- $a_1 \ldots a_i \in \mathcal{A}$ are the agents involved in the execution of s,
- $e_1 \ldots e_j \in \mathcal{E}$ represent the health modes of the required equipment,
- $n_1 \ldots n_k \in \mathcal{N}$ are the involved environment variables.

Also in this case, the variables mentioned in the range of the function f_o^{nor} must be a subset of the variables mentioned in its domain. The meaning of this model is that the execution of a plan step s, instance of a plan operator o, changes the values of the variables appearing in the range of the function f_o^{nor}. It is worth noting that the execution of a plan step can change the state of agents too.

The abnormal model of a plan operator is similar to the one already discussed in the basic framework; that is, f_o^{ab} assigns to each variable in the range of the function f_o^{nor} the undefined value $\perp$. Note that the abnormal model is applied whenever the nominal model cannot: It is sufficient that just one variable in the domain of f_o^{nor} is undefined, to make the nominal model not applicable. de Jonge et al. state that, in principle, their extended framework could accommodate action-specific health modes, however they fail in giving details on this important aspect, so it remains unclear how to model these "precise" action modes.

An important extension introduced in [14] is about the modeling of dynamic, unexpected changes that can affect the health variables of agents and equipment, and even some environment variables. These changes are modeled by borrowing some ideas from DESs literature. Changes are in fact modeled as *events* that change the values (i.e., the health modes) of some variables. Formally, an event is a triple (v, d, t), where v is a variable in $\mathcal{V}$, d is a value in the domain of v, and t is the time at which the event occurs; the meaning is very intuitive: at time t, v assumes the value d. Since different events may trigger different changes, *transition functions*, associated with variables, are also introduced in order to specify which transitions are possible. These transition functions are very close to the Mode Transition Graphs (MTGs) used in [45] and first formalized in [5]. The derivability relation that, in the basic framework is used to estimate a (partial) plan state (π', t') at time t' given another partial state (π, t) at time t (with $t < t'$), is therefore extended to deal with a set of events κ. More precisely, let (π, t) and $(\pi', t+1)$ be two consecutive partial states at time t and t+1, respectively; π' is directly derived from π if there exists a partial state (π'', t^+) such that:

1. for each $v_j \in \mathcal{V}$, $\pi''(v) = d$ if $(v_j, d, t) \in \kappa$, and $\pi''(v_j) = \pi(v_j)$, otherwise;
2. $(\pi'', t^+) \rightarrow_P (\pi', t+1)$.

The partial state π'' is an intermediate state between π and π' happening at t^+ such that $t < t^+ < t'$; that is, the time granularity of events is finer than that of states. In condition (1) π'' captures the changes that are caused by any event in κ; whereas in condition (2) the standard derivability relation is used to predict π' from π''. The soundness of the derived partial states is guaranteed by transition functions associated with variables.

The extended framework accommodates very easily the notion of plan diagnosis (i.e., the primary diagnosis). Also in this extended scenario we have a diagnostic window $[t, t']$ that individuates a precise diagnostic problem: The partial state (π', t') predicted from the partial state (π, t) assuming that all the actions behave normally is inconsistent with the partial state $obs(t')$ observed at time t'. In such a case, a qualification κ is a set of events such that the state π'_κ derived by assuming κ, $(\pi, t) \rightarrow^*_{\kappa,P} (\pi'_\kappa, t')$, is compatible with the partial, observed state *obs*(t'). The primary diagnosis is therefore a qualification $\kappa_S \subseteq \kappa$ consisting of a set of events just mentioning variables in $\mathcal{S}$. These events, in fact, represent the unexpected changes of action health modes from *normal* to *abnormal*. The secondary diagnosis, on the other hand, is a qualification $\kappa_{\mathcal{V}-S} \subseteq \kappa$ about variables that are not plan steps; i.e., variables that are in $\mathcal{V} - S$. These events represent the root causes for the abnormal actions. Also in this case qualifications can be preferred according to their predictive power, so that the notions of maxi-diagnoses and mini-maxi-diagnoses are extended in [14] to cover events.

It is important to note that, although Micalizio et al. do not use such a terminology, they actually deal with primary and secondary diagnosis: In their automaton-based approach, actions are either qualified as *on-time*, *delayed*, or *failed* (this corresponds to what de Jonge et al. call primary diagnosis). In addition, non-nominal action outcomes are explained as a combination of troublesome interactions and faults, and these explanations correspond to secondary diagnoses.

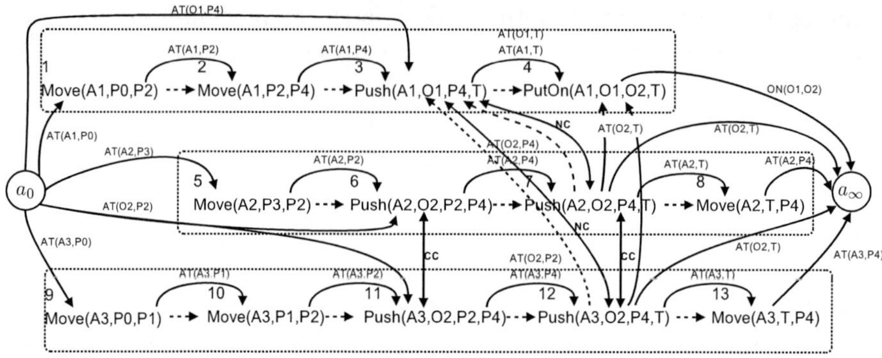

Figure 6. An example of MAP according to the relation-based framework.

6.3. Relation-Based Action Models

The relation-based framework was initially proposed in [40] and subsequently extended in [41, 42]. In this framework, the system to be diagnosed is a multi-agent plan P whose model is directly borrowed from the planning community (see e.g., [3, 12]). Given a team of agents $\mathcal{T}$, the plan P is a tuple $\langle A, E, CL, CC, NC \rangle$ such that:

- A is the set of action instances, each of which is assigned to a specific agent $i \in \mathcal{T}$. Each action $a \in A$ is characterized by a set of propositional atoms $pre(a)$ representing its preconditions, and a set of propositional atoms $\mathit{eff}(a)$ representing its effects;
- $E = \{\langle a, a' \rangle\}$, with $a, a' \in A$, is a partial-order relations over the action instances, any pair $\langle a, a' \rangle \in E$ means that the execution of a must already be finished when a' starts;
- $CL = \{\langle a, q, a' \rangle\}$, with $a, a' \in A$ and $q \in \mathit{eff}(a) \wedge q \in pre(a')$, is a set of causal links meaning that action a provides a' with the service q;
- $CC = \{\langle a, a' \rangle\}$ is a set of concurrency constraints meaning that the two actions are part of a single *joint action*; i.e., two agents have to join their efforts to obtain a result that they could not achieve on their own;
- $NC = \{\langle a, a' \rangle\}$ is a set of non-concurrency constraints, meaning that the two actions cannot be performed at the same time. This type of constraints is useful to represent the mutual access to the shared resources.

An example of plan P is shown in Figure 6. This example, borrowed from [41], shows the plan assigned to three agents $A1$, $A2$, and $A3$ that have to move the objects $O1$ and $O2$ by pushing them. Dotted boxes enclose the actions assigned to a specific agent; i.e., sub-plans. Solid edges between actions represent causal links, whereas dashed edges corresponds to precedence links. Concurrency and non-concurrency constraints are double-headed arrows labeled as CC and NC, respectively. For instance, agents $A2$ and $A3$ have to join their efforts to push object $O2$ that is assumed to be too heavy for an agent alone.

Distributed Approach. The idea proposed by Micalizio and Torasso is to exploit this rich plan model during the plan execution phase in order to detect anomalies. Their solution is fully distributed in the sense that each agent $i \in \mathcal{T}$ knows just the actions it is responsible for; dependencies existing between actions of different agents are maintained by means of links: the local plan P^i assigned to agent i is a tuple $\langle A^i, E^i, CL^i, CC^i, NC^i \rangle$, where A^i contains the actions assigned to i only, E^i and CL^i are precedence and causal links, respectively, mentioning at least one action in A^i; whereas, CC^i and NC^i are concurrency and non-concurrency constraints, respectively, mentioning exactly one in A^i and one action in A^j, with j being another agent in $\mathcal{T}$ (i.e., $j \neq i$). Thus, any link mentioning an action in A^i and another action A^j represents a dependency between the two local plans P^i and P^j that must be taken into account during the execution phase.

In this fully distributed approach, each agent is responsible for monitoring and diagnosing the actions it performs. Therefore, it is not convenient to represent the system state via a single set $\mathcal{V}$ of variables as in the resource-oriented view, because agents should communicate with one another to get such a global state at each execution time. In this framework, the system state is decomposed in agent states: each agent i has its own set $\mathcal{V}^i$ of variables that, as in the automaton-based approach, is partitioned into three subsets: END^i (endogenous variables), HLT^i (health states of the i's functionalities), and RES^i (about the shared resources). It is worth noting that, due to this distributed point of view, each agent holds a private copy for the variables encoding the status of the shared resources. Of course, all the private copies of the agents must be kept consistent throughout the execution phase.

Observability Requirements. The problem of maintaining the private variables in RES^i globally consistent, is solved via the coordinated execution of the plan actions. Coordination is achieved via direct communication among the agents. In fact, while in [14, 55] actions are performed in a *synchronized* way by all the agents, in the relation-based framework actions are performed in an *asynchronous* way; namely, each agent performs its next action as soon as it can. This requires that the agents must communicate with one another during the execution of the plan. In particular, two agents i and j have to communicate with each other in at least four cases:

- when there exists a precedence link $\langle a^i_k, a^j_h \rangle$ in E, agent i has to notify j that its action a^i_k has been completed, so a^j_h can now start (provided that j is ready to perform it);
- when there exists a causal link $\langle a^i_k, q, a^j_h \rangle$ in CL, agent i has to inform j that the service q is ready;
- when there exists a concurrency constraint $\langle a^i_k, a^j_h \rangle$ in CC, the two agents i and j have to agree to start a synchronized execution of their own actions;
- when there exists a non-concurrency constraint $\langle a^i_k, a^j_h \rangle$ in NC, i and j have to agree on serializing the two actions a^i_k and a^j_h either way.

To enable such a message passing, however, some assumptions about the system observability must be imposed. In their early work [40], Micalizio and Torasso assume that the *nominal* effects of every performed action are always observable. Relying on this assumption, they present a monitoring methodology based on a *strong commitment* policy.

The strong commitment policy, however, is based on an assumption that might appear unpractical in many applicative domains. To relax such an assumption, a minimum requirement on the system observability is introduced in [41, 42]. In the four cases above, in fact, agent i can inform j about its action a^i_k only when i can observe the effects, either nominal or abnormal, got by performing a^i_k. Action a^i_k is considered as a *target action* since it is a connection point between the i's and j's local plans. The *minimum observability requirement* imposes that at least the effects of all the target actions in the plan P must be observed by the agents performing them. On the other hand, the effects of non-target actions might not be observed at all. Relying on the minimum observability requirement, Micalizio and Torasso present in [41] a strategy for diagnosing the execution of a multi-agent plan based on a *weak-commitment* policy.

In the rest of this section we briefly review the strong and weak commitment policies; before that, however, we introduce the relation-based action models at the basis of this framework.

Relation-based Action Models. We have seen that in the resource-oriented approach by Roos et al. actions are modeled by means of functions; in particular, each action is associated with two functions: one models the normal behavior of the action, the other models the abnormal behavior mapping each variable in the range of the function to the undefined value $\perp$. Conversely, in the relation-based approach, action models are relations. Given an action instance $a^i_k \in A^i$, the model used by agent i to create expectations about the effects of a^i_k has the form:

$$\Delta(a^i_k) \subseteq (dom(v_{l_1}) \times \ldots \times dom(v_{l_m})) \times \mathcal{E} \times (dom(v_{u_1}) \ldots dom(v_{u_w})) \quad (9)$$

where $\{v_{l_1}, \ldots, v_{l_m}\} \subseteq \mathcal{V}^i$ are the variables over which the preconditions $pre(a^i_k)$ of action a^i_k are defined; $\{v_{u_1}, \ldots, v_{u_w}\} \subseteq \{v_{l_1}, \ldots, v_{l_m}\}$ is the subset of variables over which the effects $eff(a^i_k)$ of a^i_k are defined; and finally, $\mathcal{E}$ is the set of exogenous events that might happen during the execution of a^i_k, and hence might cause abnormal results. In other words, the events in $\mathcal{E}$ correspond to the faults that might affect the functionalities of agent i.

Table 2 shows an example of relation-based action model. This example, presented in [40], is about an instanced `move` action that requires agent `A` to move from its current position `P1` to location `P2`. The model represents how the variables captures in the partial state σ_t evolve in the partial state σ_{t+1}. It is worth noting that the effects of the action depend not only on the occurrence of faults, but also on the state of the agent when the action starts. For instance, the first row of the table describes the situation when no fault occurs. This is denoted with the special event ϵ included in $\mathcal{E}$. In this case, the agent reaches the expected position independently of whether the agent is empty, half-loaded, or loaded (the * symbol means "any value"). On the other hand, when a fault to the battery occurs (`f-BRY`), the impact of such an event on the action outcome depends on the initial value of the *carried* variable. Although this specific model is deterministic, it is possible to add further rows and make it non-deterministic. Indeed, works [41, 42] about the weak commitment policy (see below) deal with non-deterministic action models.

Table 2. The relation-based model for action MOVE(A,P1,P2)

	σ_t				*fault*	σ_{t+1}			
	observables		**health variables**			**observables**		**health variables**	
	position	*carried*	*power*	*mobility*		*position*	*carried*	*power*	*mobility*
1	P1	*	High	OK	ϵ	P2	*	High	OK
2	P1	EMPTY	High	OK	f-BRY	P2	EMPTY	Low	OK
3	P1	HALF	High	OK	f-BRY	P2	HALF	Low	OK
4	P1	FULL	High	OK	f-BRY	P1	FULL	Low	OK
5	P1	*	High	OK	f-MOB	P1	*	High	BK

Strong Commitment. At each step k of the execution, agent i creates expectations about the evolutions of its current action a_k^i by exploiting the relational model $\Delta(a_k^i)$, and the current belief state $\mathcal{B}_k^i$ held by the agent. Note that $\mathcal{B}_k^i$ is usually ambiguous in the sense that encodes a number of alternative agent states. Similarly to the automaton-based approach, the prediction process can be formalized in terms of Relational operators:

$$\mathcal{B}_{k+1}^i = \text{PROJECT}_{\mathcal{V}_{k+1}^i}(\mathcal{B}_k^i \text{ JOIN } \Delta(a_k^i)). \tag{10}$$

The predicted belief state $\mathcal{B}_{k+1}^i$ is subsequently refined by exploiting the observations $obs^i(k+1)$ agent i receives after the completion of action a_k^i. Also this step can be formalized in terms of Relational operators:

$$refined\mathcal{B}_{k+1}^i = \text{SELECT}_{obs^i(k+1)}\mathcal{B}_{k+1}^i. \tag{11}$$

The strong commitment policy assumes that all the nominal effects of an action, when achieved, are observed. That is, when a_k^i has not been affected by faults, the nominal effects are encoded in $obs^i(k+1)$. Thus $refined\mathcal{B}_{k+1}^i$ must hold all and only states in which the nominal effects are satisfied. In case there exists just one agent state $s \in refined\mathcal{B}_{k+1}^i$ such that the nominal effects are not satisfied, than action a_k^i is considered failed.

A diagnosis, in this case, highlights which agent functionalities behaved abnormally, and is obtained by projecting the $refined\mathcal{B}_{k+1}^i$ over the health status variables:

$$D^i = \text{PROJECT}_{HLT_{k+1}^i} refined\mathcal{B}_{k+1}^i. \tag{12}$$

Thus, here the plan diagnosis simply consists of the failed action a_k^i. The plan diagnosis is refined by means of the agent diagnosis D^i that explains the failure of a_k^i as a combination of faulty functionalities.

Note that whenever agent i detects an action failure, besides inferring the agent diagnosis, it also notifies all the other team members about the failure. An efficient protocol to propagate an action failure throughout the plan is discussed in [40]. This propagation step resembles the causal diagnosis proposed by Roos et al. in [59] through which they can estimate the plan steps that will fail in the future. In the solution proposed in [40], however, it is not necessary to use a causal theory. The structure of the plan P, with its different types of links, is sufficiently rich to accomplish the task. The problem solved in [40] is to propagate the failure in the plan structure taking into account that agents just know a portion of the plan, and that agents can be involved in mutual exchange of services.

Weak Commitment. As in the strong commitment policy, action a^i_k is considered successfully completed in the weak commitment policy only when its nominal effects hold in every state $s \in refined\mathcal{B}^i_{k+1}$. However, a^i_k is considered as failed only when its nominal effects do not hold in any state $s \in refined\mathcal{B}^i_{k+1}$. In all the intermediate situations, in which $refined\mathcal{B}^i_{k+1}$ both contains states where the nominal effects hold, and states where they do not, action a^i_k is considered as *pending*. The idea at the basis of the weak commitment policy is that the outcome of action a^i_k can be determined not only on the basis of the observations obs^i_{k+1} received just after the execution of a^i_k, but also on the basis of observations received in subsequent time instants. In fact, since the weak commitment policy just relies on the minimum observability requirement, it may happen that obs^i_{k+1} is empty when a^i_k is not a target action. On the other hand, since the effects of target actions are observable, agent i can determine, soon or later, the outcome of every action it performs (see [42]).

Dealing with pending actions, however, makes both the prediction step and the diagnosis task more complex. The prediction step, in fact, cannot just handle a single agent belief state (i.e., the last agent belief state inferred as in equation (11)), but it has to maintain a recent history or *trajectory*. Agent i needs to trace every pending action a as far as it receives sufficient information to determine either the success or failure of action a. Intuitively, a trajectory $Tr^i[0,l]$ for agent i can be seen as the sequence $\mathcal{B}^i_0 \ldots \mathcal{B}^i_l$ of the agent belief states inferred in the interval $[0,l]$.

In [41] an extension to the estimation process in formula 11 is discussed. This extension considers not only trajectories, but also agents involved in joint actions. The work [41] contributes in showing how the set $obs^i(l)$ of observations is used to filter the whole agent trajectory $Tr^i[0,l]$ built so far. The possible beneficial effect of such a pruning is to refine some belief state $\mathcal{B}^i_{k+1} \in Tr^i[0,l]$ (i.e., $k+1 \in [1,l]$) so that the outcome of a pending action a^i_k can be precisely determined. (In fact, the effects of action a^i_k are kept in the belief state $\mathcal{B}^i_{k+1}$.)

Also the diagnostic task must be adapted to deal with agent-trajectories. When agent i detects the failure of an action a^i_k performed in the interval $[0,l]$, it starts a diagnostic process. It is important to note that, as a consequence of the minimum observability requirement, agent i cannot blame other agents for the failure of one of its actions. Thus, an explanation for the failure of a^i_k must be searched within the local plan P^i. In particular, we can observe that the failure of a^i_k could either be due to a fault affecting a^i_k itself, or to a fault affecting some action a^i_h preceding a^i_k (directly or indirectly), and such that there exists a chain of causal links in CL^i that from a^i_h reaches a^i_k. This chain of causal links means that action a^i_h provides a^i_k with some service (even though indirectly). In this second case, a^i_k is a *secondary failure* caused by the *primary failure* a^i_h. In absence of further information, however, agent i cannot distinguish between primary and secondary failures. A plan diagnosis is therefore inferred by considering all the pending actions that provide services to a^i_k as failed; more formally:

$$PD^i = \{a | a \in pActs^i \cap dependsOn(a^i_k)\}, \tag{13}$$

where $pActs^i$ is the set of pending actions kept by agent i, and $dependsOn(a^i_k)$ is the subset of actions in A^i that directly or indirectly provide a^i_k with services. Each action in PD^i represents a possible explanation for the failure of action a^i_k. Note that a^i_k is trivially included in the set $dependsOn(a^i_k)$.

Each action in PD^i can be seen as a primary diagnosis, according to the terminology used by Roos et al., which can further be refined by inferring an agent diagnosis. More precisely, let a^i_h be an action in PD^i, the agent diagnosis associated with this action is given by:

$$D^i(a^i_h) = \text{PROJECT}_{HLT^i_{h+1}} Tr^i[0, l]. \tag{14}$$

That is, similarly to the strong commitment, the agent diagnosis explains why action a^i_h is failed in terms of faults occurred in some functionalities of agent i. Thus, $D^i(a^i_h)$ corresponds to the secondary diagnosis by Roos et al.

Cooperative Weak Commitment. It is important to note that the nominal or abnormal outcome of an action can be inferred not only on the basis of the observations agent i receives, but also on the basis of the information it asks to other agents.

In [42] a collaborative variant of the weak commitment policy is discussed. The basic idea is that the agents can cooperate with one another in order to gather more information about their own state. For instance, when agent i provides j with a service through a causal link $\langle a^i_k, q, a^j_h \rangle \in CL$, agent i could ask j to give some feedback about service q. According to the minimum observability requirement, in fact, if agent i cannot precisely determine the successful completion of action a^i_k, it has to assume that the action has failed. However, the observability of the system might even be below the minimum observability requirement assumed above. In this case, the plan execution would fail as soon as agent i has some uncertainty about the outcome of one of its target actions. In order to mitigate such a problem, agent i asks j to verify whether q has actually been provided or not. On the basis of the j's feedback, agent i refines its trajectory and determines the outcome of action a^i_k.

Relation-based *vs.* Resource-Oriented Methodologies. The relation-based framework by Micalizio et al. shares some similarities both with the resource-oriented view by Roos et al., and with the automaton-based approach previously proposed by the same authors. Relational actions are in fact modeled just in terms of their preconditions and effects as in the resource-oriented view, but the prediction of the next (agent) belief state is based on Relational operators as in the automaton-based solution.

There are, however, some substantial differences between the relation-based and the resource-oriented frameworks. First of all, the resource-oriented framework models actions as functions. In particular, each action instance is modeled as a pair of functions: one models the nominal, deterministic behavior of the action, the other models the abnormal behavior. The abnormal behavior, however, is modeled only weakly as it maps each variable in the range of the function in the undefined value $\perp$. On the other hand, the relation-based approach models actions as relations. Relations allows one to specifies not only the nominal behavior of one action, but also several *specific* abnormal behaviors of the action that are due to the occurrence of exogenous events (i.e., faults to the agent's functionalities). In particular, with a relational model, it is possible to model non-deterministic effects for the same fault.

Another important difference between the two methodologies is that the resource-oriented approach assumes that actions are performed synchronously by the team agents. In other words, there exists a global clock, and at each time increment of the clock each agent performs its next action. Conversely, the relation-based methodology allows the asynchronous execution of actions: an agent executes its next action as soon as the action preconditions are satisfied. This higher freedom, however, rises the problem of coordinating the agents' activities to prevent the violation of plan constraints. Coordination is achieved via direct communication among the agents. The relation-based approach imposes a minimum observability requirement to guarantee that, during the plane execution, each agent will have sufficient amount of information to coordinate with other teammates.

Finally, the resource-oriented approach infers a plan diagnosis mainly in a centralized way, whereas the diagnosis is inferred in a distributed way in the relation-based solution. However, the notion of plan diagnosis is better developed in the works by Roos et al. than in the works by Micalizio et al. The discussion about plan diagnosis presented in [55, 14] by Roos et al., in fact, considers not only the problem of qualifying a subset of plan actions as abnormal, but also to find a qualification that satisfies some preference criterion. In particular, they have shown that qualifications can be preferred on the basis of their predictive power, and that such maximally informative, minimum-maximally informative diagnoses can be inferred in polynomial time. The notion of plan diagnosis proposed by Micalizio et al. in [40, 41, 42], on the other hand, is very similar to the one proposed by Roos et al., but the minimum observability requirement they impose simplifies the diagnostic task. As an effect of the observability requirement, the problem of diagnosing the whole plan is partitioned into subproblems. Each agent needs to diagnose its own local plan only, and it can never blame another agent for the failure of the actions it is responsible for. Thus, the global plan diagnosis is partitioned into local plan diagnoses, each of which just refers to the actions of a single agent.

7. Discussion

In the last two sections we have reviewed five methodologies addressing the problem of diagnosing a MAP. The TÆMS approach by Horling et al., however, tackles diagnosis as a second-level problem. In TÆMS, in fact, diagnosis is delegated to external, specialized modules that can be added/removed to the system depending on the domain at hand. The other four methodologies instead (*social diagnosis*, *automaton-based*, *resource-oriented*, and *relation-based*), consider diagnosis as the core problem. For this reason, in this section we summarize the main features of these four methodologies in order to ease a direct comparison among them.

Table 3 highlights six important features about diagnosis of MAPs. These elements should always be considered whenever one has to decide which methodology can be adapted, or extended, in a given applicative domain. The first two features to be taken into account are concerned with the types of models one can adopt. With *system model* we refer to the way in which the team of agents is represented. In social diagnosis, the system model corresponds to a teamwork model which highlights roles and behaviors that team member can assume during their activities. Indeed, in this framework the agents' activities are not explicitly modeled but abstracted by behaviors (see *action model* entry). In the other three methodologies, the system is explicitly modeled as a plan (i.e., a MAP). Although these three methodologies differ one another in the way actions are actually modeled, they explicitly model the MAP as a set of actions and relations between actions.

Time is another important feature that characterizes the four methodologies. With *plan progress* we point to how the plan evolves over time as an effect of the activities carried on by the agents. Note that except for the relation-based approach, which is asynchronous, all the methodologies are synchronized. The granularity of the synchronization, however, depends on the specific action model being used. For instance, in social diagnosis synchronization is on the selection of behaviors, thus at a high level; whereas, in the automaton-based framework, synchronization is at a more fine-grained level as it is on the automata state transitions of the actions currently in progress.

The last three features that we report in Table 3 are about the diagnostic task. First of all, diagnosis can either be centralized or distributed (see *diagnostic inferences* entry). More importantly, the diagnostic task is focused on detecting specific types of anomalies. In social diagnosis, for instance, one is interested in discovering disagreements among the agents (i.e., coordination failures). The other three methods, instead, address directly actions. They look for mismatches between states predicted through action models and observed states. Also in this case, mismatch types depend on the specific action model being used. For instance, in the relation-based framework, the model of an action is used to estimate which effects the action execution will achieve in nominal conditions. Then, when these effects are observed to be missing, an anomaly has been detected.

Finally, we consider the different notions of diagnoses. These notions derive from the models used to describe the system and from the types of anomalies that must be detected. Social diagnosis, for instance, just identifies which beliefs held by the agents are in contrast with one another. Although this may appear at a too high level of detail, it is sufficient to explain coordination failures. On the other hand, resource-oriented, automaton-based, and relation-based frameworks identify both primary and secondary diagnoses. While primary diagnoses provide explanations in terms of abnormal actions, secondary diagnoses explain why actions behaved abnormally.

8. Conclusion

Diagnosis of MAPs is a complex task. In fact, diagnosis of MAPs can be seen as an evolution of diagnosis of component-based systems that, as seminal works on MBD ([48, 15, 13]) have already shown, is a $\mathbf{NP}\text{-}hard$ problem. Moreover, some assumptions usually made in the component-based setting might not hold when we diagnose MAPs. For example, the independence of faults does not always apply to MAPs: in some applicative domains, a single action failure might cause the failure of many other actions.

Over the last decade, different model-based methodologies to face diagnosis of MAPs have been proposed, but there is still room for further progresses. A first important contribution would be to unify in a single framework the four methodologies addressed in the previous section. For instance, the relation-based framework extends some of the notions of the resource-oriented methodology. By means of relations, in fact, one can define

Table 3. The main features characterizing model-based approaches to diagnosis of MAPs

	Social diagnosis	Resource-oriented	Automaton-based	Relation-based
System model	teamwork	plan	plan	plan
Action model	behaviors (implicit)	functions (explicit)	automata (explicit)	relations (explicit)
Plan progress	synchronized selection of behaviors	synchronized execution of actions	synchronized transition of automata states	asynchronous execution of actions
Diagnostic inferences	centralized	centralized	centralized/ distributed	distributed
Anomalies	coordination failures	inconsistencies between predicted and observed partial states of the system	troublesome interactions and action delays	missing effects
Diagnoses	contradicting agent beliefs	*primary diagnoses*: subset of actions qualified as abnormal *secondary diagnoses*: events causing abnormal actions	*primary diagnoses*: non-nominal action outcomes *secondary diagnoses*: agent diagnose expressed in terms of health status variables	*primary diagnoses*: non-nominal action outcomes *secondary diagnoses*: agent diagnose expressed in terms of health status variables

non-deterministic action models. In addition, the relation-based methodology supports the asynchronous execution of actions. These characteristics suggest that the two frameworks could be collapsed into a single framework.

On the other hand, the teamwork approach is complementary to both resource-oriented and relation-based methodologies. Thus it would interesting to study a single framework that encompass both social diagnosis and plan/agent diagnosis. In this case, the social diagnosis could be thought of as a primary diagnosis (according to the terminology by Roos and Witteveen), as it is at a high level of abstraction. Plan and agent diagnoses, instead, could be proposed to a human supervisor as refinements of the social diagnosis. The social diagnosis would detect and identify coordination failures, while plan/agent diagnosis would explain these coordination failures in terms of faults affecting the agents' devices.

Another important extension to be considered is about the temporal dimension. The approaches to the diagnosis of MAPs discussed here, in fact, do not address the time as a critical variable of the problem. Time is usually modeled as a discrete sequence of instant. At each time increment, the system is assumed to evolve synchronously; that is, all the agents are assumed to perform their next actions or to select their next behaviors at the same time. The only exception is the relation-based framework in which actions are performed asynchronously. In this case each agent has its own clock.

To be more realistic, however, the diagnostic framework should also involve temporal constraints imposed between actions. In many situations, the plan to be performed is scheduled. This means that actions are expected to start and terminate within time intervals. Two consecutive actions are therefore constrained not only by means of precedence constraints, but also by means of temporal constraints. For instance, a temporal constraint can impose that the distance between the end of the first action and the start of the second one must be lower (or greater) than a certain amount of time. In such a temporal framework, the diagnosis has to deal with a further kind of anomalies: the violation of temporal constraints.

Two approaches [54, 46] have recently been proposed to diagnose the violation of temporal constraints during the execution of MAPs. Both works exploit the formalism of the Simple Temporal Networks (STNs) [18] to detect and explain constraint violations. In [54] diagnoses consist of STN constraints that are no longer satisfied, while in [46] diagnoses are actions that took longer than expected to be completed. A possible future direction would be to combine one of these two frameworks with the plan and agent diagnoses discussed above. Temporal diagnosis could be thought of as a primary diagnosis, which could in turn be refined by the plan/agent diagnosis.

Finally, a further possible direction is about the integration between diagnosis and plan recovery. As we have noted since the introduction, in fact, diagnosis is just one of the activities that are required to make MAPs more robust. That is, systems that are, at least to some extent, tolerant of action failures. In [37, 38] a first proposal to integrate diagnosis with action failure recovery is discussed. In this approach the relation-based framework is used both to diagnose action failures, and to recover from them. The relational action models are in fact used during a conformant planning phase that aims at synthesizing on-the-fly a recovery plan. More precisely, such a recovery plan tries to restore, when possible, the nominal conditions in the functionalities of the agents. The agent diagnosis is therefore used to drive the planning process since it isolates the defective functionalities; the conformant planning then builds a plan involving self-healing actions that restore nominal the health states. Such a preliminary framework, however, just considers a single agent. Possible extensions would exploit the multi-agent setting; for instance, agents could cooperate with one another to recover from the failure of one of their teammates.

References

[1] M. Asada, P. Stone, H. Kitano, B. Werger, Y. Kuniyoshi, A. Drogoul, D. Duhaut, M. Veloso, H. Asama, and S. Suzuki. The robocup physical agent challenge: Phase - I. *Applied Artificial Intelligence*, 12(2-3), 1998.

[2] P. Baroni, G. Lamperti, P. Pogliano, and M. Zanella. Diagnosis of large active systems. *Artificial Intelligence*, 110(1):135–183, 1999.

[3] C. Boutilier and R. I. Brafman. Partial-order planning eith concurrent interacting actions. *Journal of Artificial Intelligence Research*, 14:105–136, 2001.

[4] D. Brand and P. Zafiropulo. On communicating finite-state machines. *Journal of ACM*, 30(2):323–342, 1983.

[5] V. Brusoni, L. Console, P. Terenziani, and D. Theseider Dupré. A spectrum of definitions for temporal model based diagnosis. *Artificial Intelligence*, 102:39–79, 1998.

[6] R. Bryant. Graph-based algorithms for boolean function manipulation. *IEEE Transactions on Computers*, 35(8):677–691, 1986.

[7] R. Bryant. Symbolic boolean manipulation with ordered binary-decision diagrams. *ACM Computer Surveys*, 24:293–318, 1992.

[8] R. B. Calder, J. E. Smith, A. J. Courtemanche, J. M. F. Mar, and A. Z. Ceranowicz. ModSAF behavior simulation and control. In *Proc. of the Conference on Computer Generated Forces and Behavioral Representation*, 1993.

[9] A. Cesta, G. Cortellessa, R. Rasconi, F. Pecora, M. Scopelliti, and L. Tiberio. Monitoring elderly people with the robocare domestic environment: Interaction synthesis and user evaluation. *Computational Intelligence*, 27(1):60–82, 2011.

[10] P. R. Cohen, H. J. Levesque, and H. Nous. Teamwork. *Special Issue on Cognitive Science and AI*, 25(4):487–512, 1991.

[11] L. Console and P. Torasso. A spectrum of logical definitions of model-based diagnosis. *Computational Intelligence*, 7:133–141, 1991.

[12] J. S. Cox, E. H. Durfee, and T. Bartold. A distributed framework for solving the multiagent plan coordination problem. In *Proc. of Autonomous Agents and Multi Agent Systems (AAMAS'05)*, pages 821–827, 2005.

[13] R. Davis. Diagnostic reasoning based on structure and behavior. *Artificial Intellilgence*, 24(1-3):347–410, 1984.

[14] F. de Jonge, N. Roos, and C. Witteveen. Primary and secondary diagnosis of multiagent plan execution. *Journal of Autonomous Agents and Multi-Agent Systems*, 18(2):267–294, 2009.

[15] J. de Kleer and B. Williams. Diagnosing multiple faults. *Artificial Intelligence*, 32 (1):97–130, 1987.

[16] R. Debouk, S. Lafortune, and D. Teneketzis. A coordinated decentralized architecture for failure diagnosis of discrete-event systems. In *Proc. of WODES 1998, International Workshop on Discrete Event Systems*, 1998.

[17] R. Debouk, S. Lafortune, and D. Teneketzis. Coordinated decentralized protocols for failure diagnosis of discrete event systems. *Journal of Discrete-Event Dynamic Systems*, 10(1-2):33–86, 2000.

[18] R. Dechter, I. Meiri, and J. Pearl. Temporal constraint networks. *Artificial Intelligence*, 49:61–95, 1991.

[19] K. S. Decker and V. R. Lesser. Quantitative modeling of complex environments. *International Journal of Intelligent Systems in Accounting, Finance, and Management*, 2(4):215–234, 1993.

[20] C. Dellarocas and M. Klein. An experimental evaluation of domain-independent fault-handling services in open multi-agent systems. In *Proc. International Conference on Multiagent Systems (ICMAS'00)*, pages 95–102, 2000.

[21] M. Fox and D. Long. PDDL2.1: An extension to PDDL for expressing temporal planning domains. *Journal of Artificial Intelligence Research (JAIR)*, 20:61–124, 2003.

[22] P. Fröhlich, I. de Almeida Mora, W. Nejdl, and M. Schroeder. Diagnostic agents for distributed systems. *LNAI*, 1760:173–186, 2000.

[23] S. Genc and S. Lafortune. Distributed diagnosis of discrete-event systems using petri nets. *LNCS*, 2679:316–336, 2003.

[24] W. Hamsher, L. Console, and J. de Kleer. *Readings in Model-Based Diagnosis*. Morgan Kaufmann, 1992.

[25] B. Horling and V. Benyo, B. Lesser. Using self-diagnosis to adapt organizational structures. In *Proc. Int. Conf. Autonomous Agents 01*, pages 529–536, 2001.

[26] M. Kalech. Diagnosis of coordination failures: A matrix-based approach. *Journal of Autonomous Agents and Multi-Agent Systems*, 24(1):69–103, 2012.

[27] M. Kalech and G. A. Kaminka. On the design of coordination diagnosis algorithms for teams of situated agents. *Artificial Intelligence*, 171(8–9):491–513, 1998.

[28] M. Kalech and G. A. Kaminka. Diagnosing a team of agents: Scaling-up. In *Proceedings of Autonomous Agents and Multi Agent Systems (AAMAS'05)*, pages 249–255, 2005.

[29] M. Kalech and G. A. Kaminka. Towards model-based diagnosis of coordination failures. In *National Conference of the Association for the Advancement of Artificial Intelligence (AAAI'05)*, pages 102–107, 2005.

[30] M. Kalech and G. A. Kaminka. Coordination diagnostic algorithms for teams of situated agents: Scaling up. *Computational Intelligence*, 27(3):393–421, 2011.

[31] M. Kalech, G. A. Kaminka, A. Meisels, and Y Elmaliach. Diagnosis of multi-robot coordination failures using distributed csp algorithms. In *National Conference of the Association for the Advancement of Artificial Intelligence (AAAI'06)*, pages 970–975, 2006.

[32] M. Kalech and G.A. Kaminka. On the design of social dignosis algorithms for multi-agent teams. In *Proc. International Joint Conference on Artificial Intelligence (IJCAI'03)*, pages 370–375, 2003.

[33] M. Kalech, M. Linder, and G. A. Kaminka. Matrix-based representation for coordination fault detection: a formal approach. In *Proceedings of Autonomous Agents and Multi Agent Systems (AAMAS'07)*, pages 504–511, 2007.

[34] G. A. Kaminka, D. V. Pynadath, and M. Tambe. Monitoring teams by overhearing: A multi-agent plan-recognition approach. *Journal of Artificial Intelligence Research*, 17:83–135, 2002.

[35] G. A. Kaminka and M. Tambe. Robust agent teams via socially-attentive monitoring. *Journal of Artificial Intelligence Research*, 12:105–147, 2000.

[36] A. Kraus, S. Katia, and A. Evenchik. Reaching agreements through argumentation: a logical model and implementation. *Artificial Intelligence*, 104(1-2):1–69, 1998.

[37] R. Micalizio. A distributed control loop for autonomous recovery in a multi-agent plan. In *International Joint Conference on Artificial Intelligence (IJCAI'09)*, pages 1760–1765, 2009.

[38] R. Micalizio. Action failure recovery via model-based diagnosis and conformant planning. *to appear in Computational Intelligence*, doi:10.1111/j.1467-8640.2012.00444.x.

[39] R. Micalizio and P. Torasso. On-line monitoring of plan execution: A distributed approach. *Knowledge-Based Systems*, 20(2):134–142, 2007.

[40] R. Micalizio and P. Torasso. Plan diagnosis and agent diagnosis in multi-agent systems. *LNCS*, 4733:434–446, 2007.

[41] R. Micalizio and P. Torasso. Monitoring the execution of a multi-agent plan: Dealing with partial observability. In *European Conference on Artificial Intelligence (ECAI'08)*, pages 408–412, 2008.

[42] R. Micalizio and P. Torasso. Agent cooperation for monitoring and diagnosing a map. *LNCS*, 5774:66–78, 2009.

[43] R. Micalizio, P. Torasso, and G. Torta. On-line monitoring and diagnosis of multi-agent systems: a model based approach. In *Proc. European Conference on Artificial Intelligence (ECAI'04)*, pages 848–852, 2004.

[44] R. Micalizio, P. Torasso, and G. Torta. High-level interpretation of action failures in multi-robot systems. In *Poster Proc. of Artificial Intelligence (AI), 25th SGAI International Conference on Innovative Techniques and Applications of Artificial Intelligence*, 2005.

[45] R. Micalizio, P. Torasso, and G. Torta. On-line monitoring and diagnosis of a team of service robots: a model-based approach. *AI Communications*, 19(4):313–340, 2006.

[46] R. Micalizio and G. Torta. Diagnosing delays in multi-agent plans execution. In *European Conference on Artificial Intelligence (ECAI'12)*, pages 594–599, 2012.

[47] Y. Pencolé and M.O. Cordier. A formal framework for the decentralised diagnosis of large scale discrete event systems and its application to telecommunication networks. *Artificial Intelligence*, 164:121–170, 2005.

[48] R. Reiter. A theory of diagnosis from first principles. *Artificial Intelligence*, 32 (1):57–96, 1987.

[49] N. Roos, A. ten Teije, A. Bos, and C. Witteveen. An analysis of multi-agent diagnosis. In *Proc. Autonomous Agents and Multi-Agent Systems (AAMAS02)*, pages 986–987, 2002.

[50] N. Roos, A. ten Teije, A. Bos, and C. Witteveen. Multi-agent diagnosis with spatially distributed knowledge. In *Proc. Belgian-Dutch Conference on Artificial Intelligence (BNAIC'02)*, pages 275–282, 2002.

[51] N. Roos, A. ten Teije, and C. Witteveen. A protocol for multi-agent diagnosis with spatially distributed knowledge. In *Proc. Autonomous Agents and Multi-Agent Systems (AAMAS'03)*, pages 655–661, 2003.

[52] N. Roos, A. ten Teije, and C. Witteveen. Reaching diagnostic agreement in multi-agent diagnosis. In *Proc. Autonomous Agents and Multi-Agent Systems (AAMAS04)*, pages 1256–1257, 2004.

[53] N. Roos and C. Witteveen. Diagnosis of plan structure violations. *LNAI*, 4687:157–169, 2007.

[54] N. Roos and C. Witteveen. Diagnosis of simple temporal networks. In *European Conference on Artificial Intelligence (ECAI'08)*, pages 593–597, 2008.

[55] N. Roos and C. Witteveen. Models and methods for plan diagnosis. *Journal of Autonomous Agents and Multi-Agent Systems*, 19(1):30–52, 2009.

[56] M. Sampath, R. Sengupta, S. Lafortune, K. Sinnamohideen, and D. Teneketzis. Failure diagnosis using discrete-event models. *IEEE Transactions on Control Systems Technology*, 4(2):105–124, 1996.

[57] M. Tambe. Implementing agent teams in dynamic multi-agent environments. *Applied Artificial Intelligence*, 12(2-3):189–210, 1998.

[58] G. Torta and P. Torasso. On the role of modeling causal independence for system model compilation with obdds. *AI Communications*, 20(1):17–26, 2007.

[59] C. Witteveen, N. Roos, R. van der Krogt, and M. de Weerdt. Diagnosis of single and multi-agent plans. In *Proc. Autonomous Agents and Multi-Agent Systems (AAMAS'05)*, pages 805–812, 2005.

[60] M. Yokoo, E. H. Durfee, T. Ishida, and K. Kuwabara. The distributed constraint satisfaction problem: Formalization and algorithms. *IEEE Trans. Knowledge and Data Engineering*, 10(5):673–685, 1998.

[61] M. Yokoo and K. Hirayama. Algorithms for distributed constraint satisfaction: A review. *Journal of Autonomous Agents and Multi-Agent Systems*, 32(2):185–207, 2000.

In: Advances in Computational Modeling Research
Editor: Anna Belya Kora
ISBN: 978-1-62618-065-9

Chapter 8

MODELING BEHAVIOR OF WEB USERS AS AGENTS WITH REASON AND SENTIMENT

Bosiljka Tadić*
Department of Theoretical Physics,
Jožef Stefan Institute, Ljubljana, Slovenia

Abstract

A brief recapitulation of recent progress in modeling the collective emotional behavior of Web users is given based on the author's work. A comparative relation of the models devised for the online social networks, chats and blogs, reveals suitable structure of the model and minimal requirements for the agent's attributes. Furthermore, the character of the control parameters which can be deduced from high-resolution empirical data is discussed. The potentials of the agent-based modeling of human dynamics are highlighted by few results. Some limitations of this modeling approach are also pointed out.

PACS: 02.90.+p, 89.20.-a, 89.75.Fb

Keywords: Self-organized dynamics, Online social systems, Collective emotional states, Agent-based modeling

1. Introduction

In recent years, a large amount of data from the Web permitted quantitative analysis of human online dynamics [5, 4]. It has been recognized that the Web users of a specific portal can be studied as complex dynamical system in the physics laboratory, in which the aggregate behaviors come out from the activities of individual users [9, 18, 10, 6]. The nonlinear dynamics of user's cooperation, subject to the rules and restrictions of the Web portal, is the heart of these selforganized phenomena.

However, full understanding of the underlying mechanisms requires adequate mathematical modeling of these systems. Research across distinct science disciplines, from social psychology to physics and affective computing have revealed that emotion and other

*E-mail address: Bosiljka.Tadic@ijs.si

factors affect user activity, and in this manner may affect the implied course of the whole dynamics. In particular, scientists just began to recognize that personal psychology profiles of users, their choices of topics and reasons to use the online communications, as well as the contents of the transmitted messages, define how a user will perform in a certain condition [25, 1, 15].

Taking these factors into account imposes a large obstacle for theoretical modeling. On the other hand, in recent psychology research, the emotions are studied quantitatively [3, 16, 2, 13, 21] within Russell's model of affect [14]. According to the model, two components, the arousal (degree of reactivity to a stimuli) and the valence (intrinsic attractiveness or aversiveness to a stimuli), are sufficient to determine an emotion known in common life. A further relevant prospect is an unprecedented amount of the high-resolution data from different Web portals, which enable quantitative analysis of user's own and collective behaviors. In addition, advancements of the computational power and programing tools, is of great importance when the aggregated structures of such complex units need to be simulated.

In this review, we summarize fundamental features of the agent-based models in the online systems, where the social networking among users is implicated. In accomplishing this task, here we point out the general structure of the model and requirements for the agent-based modeling across different communication systems, i.e., the online social networks, chats and blogs. Furthermore, by highlighting some results of the simulations, we demonstrate potentials and current restrictions for this modeling approach. For more results and details of the numerical implementations of the models of emotional agents, the interested reader should consult the original references by the author, specifically Refs. [11, 8] for modeling Blogs and Diggs users, [23] for the agents in online social networks, and [20] for the online Chats in the presence of Bots.

2. General Structure of the Agent-Based Model of Web Users

To describe the aggregate behavior of Web users quantitatively within the framework of agent-based modeling, three levels of the nonlinear dynamical process are considered:

- dynamics of the states of each individual user;
- the rules of action related with the individual states and interaction environment;
- the external influences and the internal fluctuations, which drive the system.

A systematic analysis of high temporal resolution empirical data from Blogs, Diggs, online Chats and online social network `MySpace` studied as complex systems in the physics laboratory, cf. Refs. [9, 10, 6, 22], a large number of regularities were found, which greatly facilitate their theoretical description. Consequently, some of the quantitative measures, presented in Fig. 1, can be used as input parameters in the agent-based modeling. In particular, these studies suggest that the principal driving force in the online human dynamics is due to *arrival of new users*. The pace that the users step into online world carries information about users' offline activity, in particular, the occurrence of circadian cycles. Other details of the user activity may also suggest specific needs of a user for the online communication. The number of new arrivals $p(t)$ at each time t, can be inferred form the empirical dataset of high temporal resolution. An example from a dataset of Diggs is shown in Fig. 1. In addition to the circadian cycles, a universal feature of all online communications that can be determined from the data is the *action delay* Δt (or *inactivity time*) for each user. The distribution $P(\Delta t)$ is broad, often with a powerlaw tail, which is characteristic of a given Web portal, see Table 1. Moreover, at each Web portal only certain subjects (posts, messages) become popular among users, leading to a broad distribution of the lifetime of the posted subjects. Similarly, a large heterogeneity occurs among the users. A power-law distribution of the number of messages per user is another relevant characteristic of an online communication system. Further observations based on the analysis of the data in which the emotional contents of messages is determined reveal interesting patterns that need to be explained by the modeling. Namely, on Blogs and similar portals where the interaction among users is indirect, mediated by the posted texts, the comments with negative emotion prevail [9, 10]. On the other hand, in the online social networks the dialogs along "friendship" links carry mostly positive emotion messages [22]. The online chats with user-to-user communications, on the other hand, have well balanced valence of messages with moderately positive overall emotion on the channel [6].

More characteristic features of the users at respective Web portal can be found and effectively used in the modeling of the system. In particular, in Blogs one can infer *innovativeness*—a fraction of new post g posted by an user, compared to the fraction of commented posts of other users. Furthermore, herding and selectivity, i.e. how often a user gives attention to an old popular post, can be measured by a parameter $\mu(T_0)$, where T_0 is a window when new posts are "exposed" by the system moderators. In the online chats, there are other parameters that may designate each user, like the number of messages in a given time window. Taking the average over all users in the dataset, the robust distributions of the defined parameters emerge, which can be considered as quantitative indicators of the system. Thus, in modeling users on a given Web portal, the *attributes* of the agents should be selected from the corresponding distributions.

In general, modeling an agent in these online communication systems, one should consider a minimal set of its attributes as follows:

$$A[id, status; (state\ variables); (activity\ profile); (connections)]. \tag{1}$$

Here, a unique id and $status \in [active, inactive]$ is known. The $state\ variables$ in the present case are the agent's emotional state arousal and valence (a, v), which fluctuate in time (see below). The $activity\ profile$ can be, for instance, taken from the distributions of innovativeness $g \in P(g)$, total number of messages $m \in P(m)$, or user's inclination to post a negative comment α. These quantities can be determined from the empirical data when the texts of messages are annotated against the emotional content, e.g., by `SentiStrength` classifier [21, 12]. These parameters can be fixed for a given agent and kept quenched during the simulation [11, 23]. The $connections$ on the network are lists of neighbors on a fixed network, as friendship links in the online social networks, or lists of neighbors established during the dynamics in the chats and blogging. Note that the connections are specific to each agent on the network and can vary in time.

The emotional state of each agent, measured by the actual values of the arousal $a_i(t) \in [0,1]$ and valence $v_i(t) \in [-1,+1]$, are subject of a dynamical model, which, in principle needs some justification from psychology research. In the absence of detailed information,

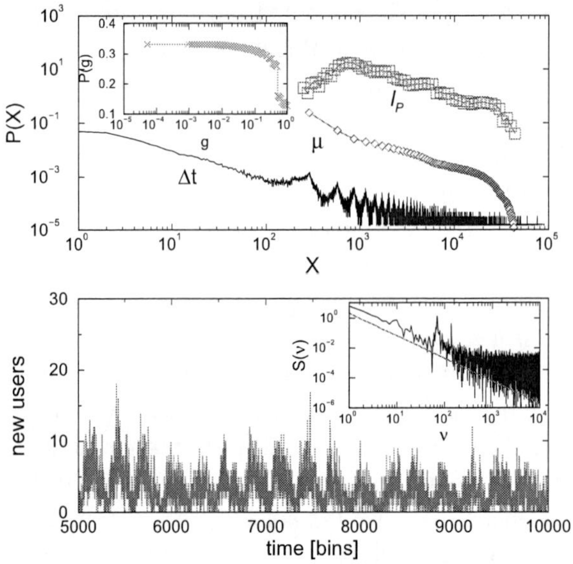

Figure 1. Some quantitative measures inferred from the empirical data of Diggs, which are used as parameters in the model: (bottom) New-users time series $p(t)$ with 5 minutes resolution, and its power spectrum in the inset; (top) Distributions of the user delay times $P(\Delta t)$ and innovativeness $P(g)$, and the lifetime of posts $P(\ell_P)$, and the probability $\mu(T_0)$.

a model with polynomial nonlinearity was suggested [17]. Here, we adapt the following evolution equations for the arousal and valence of an agent i situated on a network:

$$a_i(t+1) = (1-\gamma_a)a_i(t) + \delta_{i,t}\left[\frac{h_i^a(t)+qh_{mf}^a(t)}{1+q}\left(d_1+d_2(a_i(t)-a_i(t)^2)\right)\right](1-a_i(t)), \quad (2)$$

$$v_i(t+1) = (1-\gamma_v)v_i(t) + \delta_{i,t}\left[\frac{h_i^v(t)+qh_{mf}^v(t)}{1+q}\left(c_1+c_2(v_i(t)-v_i(t)^3)\right)\right](1-|v_i(t)|). \quad (3)$$

Here, the relaxation term with the rate γ applies constantly. Whereas, the contributions from the nonlinear terms depend on the events that take place at the network at moment t, and may or may not affect the agent i (indicated by the Kronecker symbol $\delta_{i,t}$). The genuine advantage of this model is in the feedback fields that provide agent's arousal and valence. They take into account current changes in the surroundings of a particular agent i on the network, $h_i^{a,v}(t)$, and in the entire system, $h_{mf}^{a,v}(t)$, encountered by all agents. A portion q of the mean-field contribution depends on the openness of all information to the users, which depends on the system analyzed. Besides, differences in the communication rules at Blogs, Chats or social networks, define how these fields can be calculated and how they develop in time. For example, in the online social networks only the messages directed to the agent i from its neighbors can give raise its fields as [23]: $h_i^z(t) = \frac{\sum_j \sum_{m\in M_{ji}} \theta(t,t_m) z_j(t_m) W_{ji} e^{-\gamma^h(t_{ji}^{lm}-t_m)}}{\sum_j \sum_{m\in M_{ji}} \theta(t,t_m) W_{ji} e^{-\gamma^h(t_{ji}^{lm}-t_m)}} e^{-\gamma^h(t-t_{ji}^{lm})}$, where M_{ij} is the message stream from the agent j to i. Whereas, in the case of Blogs, such fields arise from the comments of other agents on the posts to which the agent i is connected [11, 8]. Likewise, in the case of chats with a recognized agent, these fields comprised of the

Table 1. A comparison of three-level ingredients in the agent-based modeling of three online communication systems. Abbreviations: *b.d.w.* and *m.d.w.* indicate the bipartite and monopartite directed weighted networks

Level	Description	Blogs & Diggs	MySpace	Online Chats
Individual	Relaxation	$\gamma_a = \gamma_v$	$\gamma_a = \gamma_v \neq \gamma_h$	$\gamma_a = \gamma_v$
	Action profile	$g \in P(g)$; $\mu(T_0)$	ε	$m \in P(m), g \in P(g)$
	Delay	$\Delta t \in P(\Delta t)$	$\Delta t \in P_{ms}(\Delta t)$	$\Delta t \in P(\Delta t)$
Environment	Network	evolving, b.d.w.	fixed, m.d.w.	evolving, m.d.w.
	Post-lifetime	$t_P \in P(t_P)$	–	–
	Exp.window	T_0	–	T_0
	Visibility	q	(nnn)	q
	Reactivity	–	a_0	a_0
External	New user noise	$p(t)$	$p(t)$	$p(t)$
	Input noise	–	p_0	–

messages directed to the agent within the active time window T_0, which is channel specific [20]. The mean fields are constituted in a similar way, but with taking the average over the network (or its active part). For mathematical compatibility of the maps (2)-(3) it is critical to have normalized fields properly at each time step. More explicit account of the model rules and numerical implementations for respectively three online systems considered can be found in the above references [11, 8, 20]. Here, a comparison of the principal features of the present models is given in Table 1.

3. What Can Be Accomplished by Agent-Based Modeling?

For describing the aggregate behaviors of the agents, a number of measures, *different from the ones which are employed as input parameters* are computed in the simulations. Here, with few results we illustrate a wide range of potentials of the agent-based modeling.

Emergence of a network in Blogs and Chats: In Fig. 2 a representative network that appears via blogging with the emotional agents is displayed. A comprehensive analysis of such networks [8] reveals that they have much the same topology as the networks deduced from authentic data, when the inherent parameters are employed. The communities of users on Blogs occur in connection with particular popular posts. In the model, the community structure can be tuned by modifying the fraction q of the knowledge accessible to all agents.

Correlations in message streams: In the implementation of our models, we hold information of each separate message (comment) together with the emotional content that it carries, as well as every activity of the agents over time. Of this degree resolution of the simulated data permits us to calculate and analyse distinct types of time series. For the illustration, in Fig. 3 the series of emotion carrying messages are shown. Their power spectrums show that the streams of messages carrying negative emotions have the degree of correlation different from the ones with positive emotion.

Avalanching and self-organization: Outbursts (avalanches) of the emotional messages occur both in original online communications and in simulations within suitable agent-based models. Nevertheless, the underlying mechanisms can be examined exclusively

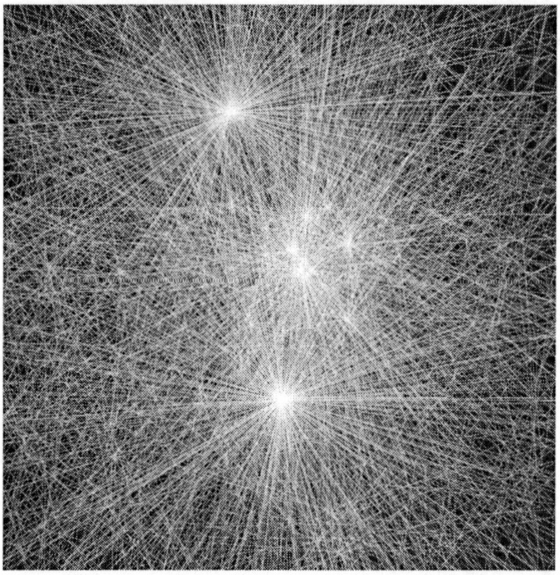

Figure 2. Emergence of popular posts in the blogging network with emotional agents.

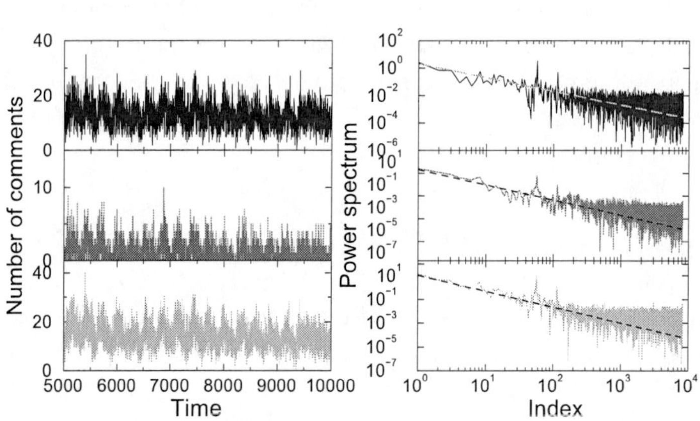

Figure 3. Time-series and power spectrum of the messages carrying negative and positive emotion, top and middle, and all messages, bottom, from agent-based model of Blogs [11].

within the models. A quantitative measure is obtained by analysing the distributions of size and duration of the avalanches and the occurrences with the emotional contents, cf. Fig. 4. Then using the theory of selforganized criticality [19, 7], one can size up the collective dynamics and perhaps identify a dynamical critical state, which is related with a parameter [10]. Moreover, investigation of the shape of the avalanches provides valuable information about the character of the selforganized process and can indicate the possible presence of a coherent noise.

Patterns of emotion flow: In the agent-based modeling, the flow of emotion can be controlled both on the network and in the phase space of the emotion variables as demonstrated in Fig. 5. The maximum-flow spanning tree of the network of agents, where each agent is connected by its best link, e.g. the largest number of messages or maximum arousal of all

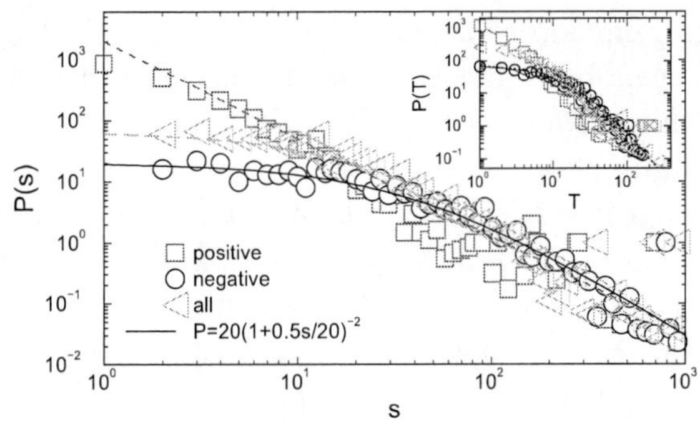

Figure 4. Emotion-carrying avalanches of comments in the agent-based model of Blogs.

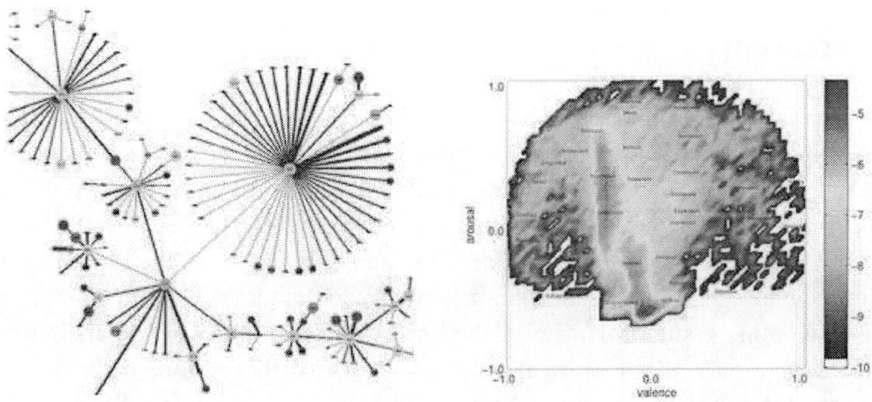

Figure 5. A part of the maximum flow tree in the positive emotion propagation on a social network (left) and a pattern with marked emotions by agents involved in blogging (right).

messages, is built from the simulated data. It renders a pattern of most significant paths for the flow of messages with a selected attribute. Furthermore, keeping track of all messages and their contents, the simulation reveals what emotions (by their arousal and valence components) of the agents were, in fact, involved in the process.

3.1. Potentials and Limitations

Owing to detailed information about the system within the agent-based model, the simulated data permit for calculating a comprehensive set of measures of the individual and group states of the agents. Above some of these measures are quoted. Thorough analysis can be found in the associated references. On the condition that the rules and the parameters are chosen from a specified empirical dataset, the simulation results can be estimated comparatively with the ones of the original system to validate the model. Likewise, by investigating the cohesion of distinct measures of the aggregate behaviors within the theory of complex dynamical systems, the simulated data can give an insight into the mechanisms that have been effective in the empirical dataset. The real significance of the agent-based modeling, however, is in the experimenting with the system, i.e., by modifying the parameters in a wide range, changing circumstances and driving modes in which agents are communicating. In this way, the agent-based model can predict the probable outcome (quantified by changed indicators) in this nonlinear system and identify the pertinent parameters that can affect the direction of the events.

Presently a weak point in this procedure is the absence of psychology consensus on which the mathematical model can be defined, i.e., to replace the plausible Eqs. (2) and (3). It should provide sufficient information about co-evolution of the components of emotion for an isolated user. Ideally, we would seek a quantitative model in analogy to natural sciences. For example, research in the chemical kinetics of gene expressions lead to two coupled equations for the evolution of the concentrations of mRNA and proteins in the process [24].

4. Conclusion

The dynamics of user behaviors in the online world can be elucidated within agent-based modeling by agents with human attributes—reasons, choices and emotions. Within our modeling approach, the emotion components evolve according to specified differential equations. Whereas the remaining attributes are obtained in an *implicit form* from the patterns of user activity in the empirical data. The general structure of the model with three layers—individual agents, interaction on a network and external influence, is applicable across diverse online systems. However, the intricate rules and the input parameters need to be adjusted to a particular empirical system. Favorably, they can be deduced from the available online data. At the same time, this reinforces a predictive use of the model. The requirement of further inputs from psychology research is pointed. Especially, the components of an emotion should coevolve in a manner, reasonably representing a psychology profile of a user. Without this, even though the Eqs. (2) and (3) might be formally correct, they do not gain credibility in the psychology society. The mathematical structure of the models can readily include additional knowledge from psychology research. Thus, the agent-based models embody robust tools at the interface between social psychology and affective computing.

Acknowledgment

This work was supported by the program P1-0044 by the Research agency of the Republic of Slovenia and from the European Community's program FP7-ICT-2008-3 under grant n° 231323. I would like to thank Milovan Šuvakov and Marija Mitrović for fruitful collaboration on the works in Refs. [23, 20, 11, 8].

References

[1] Y. Amichai-Hamburger and G. Vinitzky, "Social network use and personality," *Computers in Human Behavior*, vol. 26, no. 6, pp. 1289 – 1295, 2010, online Interactivity: Role of Technology in Behavior Change. [Online]. Available: http://www.sciencedirect.com/science/article/pii/S0747563210000580.

[2] R. Calvo and S. D'Mello, "Affect detection: An interdisciplinary review of models, methods, and their applications," *Affective Computing, IEEE Transactions on*, vol. 1, no. 1, pp. 18 –37, 2010.

[3] J. A. Coan and J. J. B. Allen, Eds., *The Handbook of Emotion Elicitation and Assessment.* Oxford University Press Series in Affective Science, 2007.

[4] J. Giles, "Social science lines up its biggest challenges," *Nature*, vol. 470, pp. 18–19, 2011.

[5] M. Giles, "A world of connections-A special report on social networking," *The Economist*, p. 16, 2010. [Online]. Available: http://www.economist.com/node/15351002.

[6] V. Gligorijević, M. Skowron, and B. Tadić, "Structure and stability of online chat networks built on emotion-carrying links," *Physica A: Statistical Mechanics and its Applications*, vol. 392, no. 3, pp. 538 – 543, 2013. [Online]. Available: http://www.sciencedirect.com/science/article/pii/S0378437112008837.

[7] H. J. Jensen, *Self-Organized Criticality: Emergent Complex Behavior in Physical and Biological Systems (Cambridge Lecture Notes in Physics).* Cambridge University Press, 1998.

[8] M. Mitrović and B. Tadić, "Dynamics of bloggers' communities: Bipartite networks from empirical data and agent-based modeling," *Physica A: Statistical Mechanics and its Applications*, vol. 391, no. 21, pp. 5264 – 5278, 2012. [Online]. Available: http://www.sciencedirect.com/science/article/pii/S0378437112004918.

[9] M. Mitrović, G. Paltoglou, and B. Tadić, "Networks and emotion-driven user communities at popular blogs," *European Physical Journal B*, vol. 77, pp. 597–609, 2010.

[10] M. Mitrović, G. Paltoglou, and B. Tadić, "Quantitative analysis of bloggers' collective behavior powered by emotions," *Journal of Statistical Mechanics: Theory and Experiment*, vol. 2011, no. 02, pp. P02 005+, 2011. [Online]. Available: http://dx.doi.org/10.1088/1742-5468/2011/02/P02005.

[11] M. Mitrović and B. Tadić, "*Patterns of emotional blogging and emergence of communities: Agent-based model on bipartite networks*," 2011. [Online]. Available: http://arxiv.org/abs/1110.5057.

[12] G. Paltoglou and M. Thelwall, "Seeing stars of valence and arousal in blog posts," *IEEE Transactions on Affective Computing*, vol. 99, pp.1+, 2012.

[13] G. Paltoglou, M. Thelwall, and K. Buckely, " Online textual communcation annotated with grades of emotion strength." in *Proceedings of the Third International Workshop on EMOTION (satellite of LREC): Corpora for research on emotion and affect*, pp. 25–30, 2010.

[14] J. A. Russell, "A circumplex model of affect," *Journal of Personality and Social Psychology*, vol. 39, pp. 1161–1178, 1980. [Online]. Available: http://dx.doi.org/doi/10.1037/h0077714.

[15] T. Ryan and S. Xsenos, "Who uses facebook? an investigation into the relationship between the big five, shyness, narcissism, loneliness, and facebook usage," *Computers in Human Behavior*, vol. 27, pp. 1658–1664, 2011. [Online]. Available: http://dx.doi.org/doi/10.1016/j.chb.2011.02.004.

[16] K. Scherer, "What are emotions? And how can they be measured?" *Social Science Information*, vol. 44, no. 4, pp. 695–729, 2005. [Online]. Available: http://ssi.sagepub.com/cgi/content/abstract/44/4/695.

[17] F. Schweitzer and D. Garcia, "An agent-based model of collective emotions in online communities," *Eur. Phys. J. B*, vol. 77, no. 4, pp. 533–545, 2010.

[18] M. Szell and S. Thurner, "Measuring social dynamics in a massive multiplayer online game," *Social networks*, vol. 39, pp. 313–329, 2010.

[19] B. Tadić and D. Dhar, "Emergent spatial structures in critical sandpiles," *Phys. Rev. Lett.*, vol. 79, pp. 1519–1522, 1997.

[20] B. Tadić and M. Šuvakov, "Agent-based model of online chats with Bots: Controlling avalanches of emotional messages," 2012. [Online]. Available: http://www-f1.ijs.si/~tadic/cv_ordered.html#ABM.

[21] M. Thelwall, K. Buckley, G. Paltoglou, D. Cai, and A. Kappas, "Sentiment strength detection in short informal text," *J. Am. Soc. Inf. Sci. Technol.*, vol. 61, pp. 2544–2558, 2010. [Online]. Available: http://dx.doi.org/10.1002/asi.v61:12.

[22] M. Šuvakov, M. Mitrović, V. Gligorijević, and B. Tadić, "How the online social networks are used: Dialogs-based structure of MySpace," *Journal of the Royal Society Interface*, vol.10, 2013. doi: 10.1098/rsif.2012.0819.

[23] M. Šuvakov, D. Garcia, F. Schweitzer, and B. Tadić, "Agent-based simulations of emotion spreading in online social networks," 2012. [Online]. Available: http://arxiv.org/abs/1205.6278.

[24] S. Widder, J. Schicho, and P. Schuster, "Dynamic patterns of gene regulation I: Simple two-gene systems," *J. Theor. Biol.*, vol. 246(3), pp. 395–419, 2007.

[25] T. Yarkoni, "Personality in 100.000 words: A large-scale analysis of personality and word use among bloggers," *Journal of research in personality*, vol. 44, pp. 363–373, 2010.

INDEX

D

E

F

G

H

I

K

L

M

N

O

P

T